D0872039

Farm Management in Africa

THE PRINCIPLES OF PRODUCTION
AND PLANNING

MARTIN UPTON

London

OXFORD UNIVERSITY PRESS

1973

Oxford University Press, Ely House, London W.1

GLASGOW NEW YORK TORONTO MELBOURNE WELLINGTON
CAPE TOWN IBADAN NAIROBI DAR ES SALAAM LUSAKA ADDIS ABABA
DELHI BOMBAY CALCUTTA MADRAS KARACHI LAHORE DACCA
KUALA LUMPUR SINGAPORE HONG KONG TOKYO

Hardback edition ISBN 0 19 215646 2

Paperback edition ISBN 0 19 215647 0

© Oxford University Press 1973

Figures drawn by Oxford Illustrators Limited

Printed in Great Britain by Richard Clay (The Chaucer Press) Ltd.,
Bungay, Suffolk.

Preface

This book is intended as an introduction to the principles of production economics, farm management and planning for students whose interest lies mainly in the agriculture of African countries. For such students most books on farm management written in Europe and America are unsatisfactory not so much because their examples have little relevance to African conditions (though winter feeding of dairy cows, choice of combine harvesters or understanding of balance sheets may be of little interest in Africa) but because of the general approach. Farm management in Western Europe and America has always been concerned, in the main, with individual farm accounting, business analysis and planning. In much of Africa this approach is not feasible. Farms are too small and farm management advisers too few to justify the analysis and planning of individual holdings. Instead a group approach is needed; the farm management adviser must attempt to analyse and understand the management problems of groups of similar farmers; likewise he must plan and evaluate the effects of a change on particular groups of farms rather than on individual holdings.

The farm management adviser is an important link in the chains of communication between farmers and government policy makers on the one hand and between farmers and technical research workers on the other. Thus he has two responsibilities; one is to investigate and diagnose the weaknesses of existing farm systems so that he can make recommendations for agricultural production policies and give directions for technical research; the other is to evaluate the likely effects of new policies and technical innovations on patterns of farming.

To carry out these responsibilities effectively a farm management specialist needs to know not only how to investigate farm problems and plan improvements but also the basic theory of production and the factors influencing the distribution and use of resources. The book is therefore divided into four parts. Part I deals with the economic theory of production and shows how resources should be allocated in order to maximize profits. It ends with a discussion of how the theory should be modified to take account of varying prices and risk. Part 2 deals with the distribution and use of the resources of land, labour, capital and management. Part 3 is concerned with methods of farm management investigations and Part 4 with farm planning. There are mathematical appendices to Chapters 2, 4 and 8 which may be omitted on a first reading. However, the mathematics involved does not go beyond elementary algebra and these appendices should improve the reader's understanding of the preceding chapters as well as material in Chapters 13 and 16.

Warmest thanks are expressed to the following for their criticisms of earlier drafts of the book. I alone am responsible for the errors.

Professor Q. B. O. Anthonio
Mr. Gordon Bridger
Dr. Michael P. Collinson
Dr. Graham Dalton
Mr. Gordon Gemmil
Dr. Michael Jawetz
Miss Edith H. Whetham

M.U.

Stratfield Saye
1972

Contents

uncertainty and subsistence farming—risk precautions in commercial
farming

Part 2 RESOURCE USE

The environment—population density—natural resources—markets and
communications—land tenure

The measurement of labour—the productivity of family labour—hired
labour and opportunity costs—hours worked and leisure preference—the
seasonality of farm labour requirements—increasing labour productivity—
incentives—improvement of working methods

Definition and measurement—investment and depreciation—return on
capital and risk discounts—the cost of saving and the supply of capital—
new forms of capital—credit—interest and security for loans—borrowing
and risk—private loans and moneylenders—government and cooperative
credit—supervised credit and hiring

Mathematical appendix
Compound interest, the cost of waiting—annuity—discounting—
discounting an annuity—land values—the internal rate of return—the oil
palm example

The productivity of management—attitudes and objectives—managerial
ability—influence of environment and background

Part 3 INVESTIGATION AND ANALYSIS

Farm management data and their uses—the data needed—existing
sources—technical experiments—case studies and unit farms

Their advantages and costs—the average farm—sampling—organization
—measurement

Tables

APPENDIX

Figures

Part 1 THE THEORY OF PRODUCTION

---

1 The scope of farm management

PRODUCTIVE RESOURCES

Farmers are producers of food and other useful commodities from plants and animals. The growth of plants and animals goes on in nature without participation by man, but farmers take control of this growth, stimulate it and use it for their own purposes. For this process of production they need a supply of productive resources such as land, seeds, breeding stock, human skill and effort, tools and machines. These productive resources, also known as factors of production, are usefully grouped into four main categories: (1) natural resources, (2) labour, (3) capital, (4) management.

The first category consists of those gifts of nature which are not the result of human effort, including land, water and local climate. Land in this sense is defined as the 'original and indestructible properties of the soil', but it must be admitted that the distinction between natural resources and the other categories is arbitrary. For land has often been made productive as a result of human effort in clearing, cultivating, manuring and perhaps irrigating and draining; the supply of water has often been improved by the building of dams or storage tanks or the construction of canals. These results of past efforts are a form of capital and should be included in the third category.

The term labour describes the effort of human beings, including that of farmers, their families and the hired workers. Although many forms of industrial production and even pig and poultry keeping require very little land, labour is needed for every kind of production. Labour can become more productive as a result of time and effort devoted to training and education. Here again the skill and knowledge acquired may be classified as capital.

Capital represents resources which are the result of past human effort. This category includes a wide diversity of items ranging from durable capital such as buildings, dams, roads and machinery to stocks of materials like seed or fertilizer which may be used up within a single season. In everyday speech capital is often confused with money savings, but it is important to distinguish between them for money is not itself a productive resource. It only becomes productive when it is used to buy physical items of capital.

These resources of land, labour and capital are organized into productive units which, in agriculture, are called farms. In parts of Africa, the word 'farm' is used for a single plot of land but in 'farm management' it means all the productive resources under the control of one farmer or farm family so it may include several plots of land and possibly several kinds of livestock. The quantities of productive resources or the services they provide which are used on a farm over a given production period may be called inputs to distinguish them from the outputs produced which are available either for consumption, sale or use as capital inputs in future production periods.

Management describes the function of taking decisions about how land, labour and capital resources should be used and carrying out these decisions. All production implies the taking of some risk, since decisions are made and inputs committed on the basis of expected yields and prices. Actual outcomes may be better or worse than projected because of either bad luck or bad decisions. In some forms of business, risk taking may be treated as a separate, and fifth, category of productive resources but in farming the farmer himself takes the risks and bears the consequences of his own decisions so risk taking is included with the management factor.

Typically African farmers control few productive resources by the standards of Western Europe and America where many of the theories of farm management were developed. In particular, capital is limited so that few machines are used and, with the exception of some cash-cropping regions, not much labour is hired. Hence the land farmed is limited to the amount that can be cultivated by the farm family using a few, simple hand tools.

By contrast large commercial farms based on foreign-owned capital and known as plantations were established in many parts of Africa in the last fifty years, particularly in Tanzania, the Sudan and the Congo. The main objectives in establishing these plantations were the provision of raw materials for foreign processors and the earning of profits for the foreign owners who provided the capital and the management and took the risks of success or failure in a commercial business. More recently, the African governments of newly independent countries have both taken over the existing plantations and also established new ones to increase agricultural production. Even larger farming units have been tried in the Kongwa groundnut scheme in Tanzania, the Niger agricultural project in Nigeria and the state farms of Ghana. Settlement schemes involve the resettling of a large number of farmers into one consolidated planting area. For purposes of land cultivation and crop processing these may be operated as single units but many of the management decisions are taken by the individual settlers so these schemes have characteristics of both large and small farms.

There is a wide difference in size between these very large units established with government or foreign-owned capital and the indigenous small farms. However, there are local farmers in most areas, though more particularly where a major cash crop can be grown, who are buying or hiring more productive resources to produce more output than on the typical small farm described above.

Another important characteristic of most African farmers is that they still grow food crops for their own consumption or subsistence even where cash crops are also grown for sale, although today we rarely find pure subsistence farming where the output is used entirely for the consumption of the farm family. This does not necessarily imply that African farm families are only just managing to subsist, though this may be true in some parts of the continent. Some quite wealthy farmers still choose to grow their own food. This means that farm management is closely related with home management and consumption economics; decisions on the farm directly influence the consumption alternatives open to the family. A decision to grow manioc (cassava) rather than yams need not be influenced by the family's consumption preferences on a purely commercial farm but on a subsistence farm this decision determines what the family will eat.

MANAGEMENT DECISIONS

On the plantations and large commercial farms, as in most industries, the workers who provide physical labour and the managers who take decisions are separate individuals. Small farmers, on the other hand, are both workers and managers, faced with the problems of organizing the labour of themselves and their families in conjunction with other inputs. Every agricultural producer, whether he is farming 2 hectares for subsistence, 200 hectares for profit or 2000 hectares of a state farm, is faced with the same broad groups of decisions about the allocation of his productive resources.

Firstly the farmer must decide what to produce from among the alternatives open to him. Will his resources be best employed in growing cotton or food crops, in producing pigs or eggs? Should he devote some inputs to constructing grain stores or irrigation canals?

Secondly, he must decide how much to produce, both in total and of each of the separate products, for, with a given level of inputs, expansion of one activity must involve the contraction of another. To produce more cotton on a particular farm, it may be necessary to reduce the area of food crops. Alternatively inputs may be increased, possibly involving the farmer and his family in more work. Then what proportion of the family's time should be devoted to leisure or participation in community organizations and activities?

Thirdly, there are decisions regarding the method of production. Should the farmer rely on shifting cultivation to maintain soil fertility or should he use manures and fertilizer and crop the land continuously? Should cultivations be carried out entirely by hand tools or should a tractor and its implements be used, so that a larger area can be cultivated? What kind of implements should be used? How frequently should the crops be weeded? Should livestock be penned or housed and fed by hand? What combinations of ingredients should be included in their rations?

These are examples of the decisions farmers must take in organizing their farms, and the purpose of farm management theory is to provide a rational basis for solving these problems. They can be summarized under three headings:

(1) what to produce;
(2) how much to produce;
(3) what method of production to use.

In addition, commercial farmers are faced with the problems of selling their output and of buying their inputs, but these questions of marketing are only incidental to the subject matter of this book. They are dealt with more fully in Edith H. Whetham's *Agricultural marketing in Africa*, London, Oxford University Press, 1972.

THE TRADITIONAL SOLUTION

Farmers may take their decisions as managers by following tradition, that is, they grow the same crops in the same quantities and by the same methods as they have always done, and their fathers did before them. In some circumstances tradition may be the safest guide to a solution. Traditional methods have been developed by generations of practical farmers to suit local conditions, and their experience has shown that people can subsist, at least, by growing the customary crops by traditional methods. By contrast new crops and new methods, even though they have been tested several years by research workers, may not succeed under local conditions, so innovators take considerable risks in trying out new ideas.

Farmers bound by culture and tradition to the customary pattern of cultivation may make few decisions on rational grounds since they do not consider alternatives. But today, population pressure in some areas, and more generally a greater awareness of wider opportunities for improving their standard of living, are causing farmers to consider and accept change. Science and technology are providing new varieties of crops, new methods of producing them and new methods of feeding animals and of controlling animal diseases, so it is possible for progressive farmers to increase the output from their limited resources of land, labour and capital. At the same time there are new opportunities for spending on bicycles, radios, watches, expensive clothes and new imported foods, besides school fees and doctors' bills. Economic development and rural education are making even the traditional farmers more aware of possible alternatives in their use of resources, and thus encouraging them to consider changes.

THE TECHNICAL SOLUTION

It is sometimes suggested that the solutions to the decisions about what to produce, how much to produce and what methods of production to use can be found through agricultural science and technology. Thus the question of what to produce might be decided by considering the soils, natural vegetation and local climate. The crops or livestock best suited physiologically to these conditions might then be recommended as the answer to the question 'what to produce'. This is very often the approach taken in 'land-use planning' by geographers and soil scientists.

Secondly the question of how much to produce might be decided by producing the highest possible yield. If we want to know how many times a crop should be weeded or irrigated, the answer might then be the number of times likely to lead to the highest yield per hectare, a point sometimes known as the 'technical optimum'. This is a particularly harmful recommendation in many parts of Africa where land is plentiful and yield per man is far more important than yield per hectare. The technical optimum level of feeding livestock is where growth rate or the yield of milk or eggs is at a maximum.

The answer to the third question, what method of production to use, may depend upon the special field of interest of the technical adviser. The agricultural engineer might recommend mechanization in order to save manual work and cultivate as much land as possible; the plant breeder might seek to increase production by introducing new seeds; the veterinarian is interested in methods of livestock keeping which minimize disease risk. It is suggested by one agricultural economist who has worked in both East and West Africa that 'when a country is poor, there is always a danger that a technician who can see how to make *some* improvement will be given too free a hand'.[1]

In fact, few agricultural scientists would take such narrow views of the decisions about the use and management of productive resources, since most of them are aware that farmers' attitudes and objectives must be taken into account. The technical solutions suggested above provide the range of alternative choices open to farmers, from which they choose those courses of action which seem most likely to achieve their objectives.

THE ECONOMISTS' SOLUTION

The contribution of economics to the decision-making process lies in 'counting the cost'. The word cost is used here in its broadest sense to mean not only expenditure of money but also the sacrifice of leisure, of food for seed or of anything else which is valued by the farm family. All farm management problems can then be considered in relation to the expected benefits and the costs incurred. The problem of what to produce is decided by choosing the combination of crop and livestock products which is likely to leave the largest surplus of benefits over costs. The problem of

[1] Martin, A. (1957) A note on research methodology and peasant agriculture, *Farm Economist*, vol. 8, no. 9, p. 47.

how much to produce is decided by increasing output until the extra benefits no longer exceed the extra costs. The problem of what method of production to use is decided by choosing the method with the lowest cost per unit of product.

This approach to decision making is based on the assumption that farmers grow crops and keep livestock for the satisfaction of their personal wants and those of their families, either directly in subsistence farming or indirectly when products are sold and the money used to buy what they want. Satisfaction is increased by the benefits of farm output and decreased by the costs of sacrificing food, leisure or money or taking risks. The rational way of choosing between alternative farming practices is to make the surplus of benefits over costs as large as possible, that is, to maximize satisfaction.

Of course, comparison of benefits and costs is only possible if technical information is available on the physical relationships between inputs of land, labour and capital and expected outputs for each alternative open to the farmer. Estimates of these technical relationships are sometimes called input–output data. To take risk into account it is necessary to know not only average or expected inputs and outputs but also the variation that may occur from year to year. For traditional products and farming practices farmers know, from past experience, how much land, labour, seed and other inputs they are likely to need and what yields they can expect. They are also aware of how much these quantities may vary. A farm management adviser must be prepared to collect this information from farmers in order to understand the problems they face and to assess the effects of new techniques on existing patterns of farming. Information on the new techniques must originate from agricultural scientists and technologists, although farm management specialists may be called upon to try out innovations under farm conditions before they are widely recommended.

The assumption that farmers manage their farms so as to maximize their satisfactions implies that they are capable of evaluating the satisfaction they get from different inputs and outputs or, at any rate, that they can arrange the alternatives open to them in some order of preference. These subjective values and preferences vary from one person to another so different individuals will choose to manage their farms differently. For this reason, advisers in farm management cannot take decisions for farmers, but farmers can be helped to think rationally about the new techniques and new opportunities now available to them. Furthermore in some circumstances advisers would be justified in using objective measures of value. For instance in a community barely managing to subsist, the satisfaction that every farmer gets from his outputs of crops and livestock must be closely related to their nutritional values, which can be measured scientifically. Alternatively where trading is common, in what we may call an 'exchange economy', prices, which are money values, may be used.

Where money or other objective measures are used the value of product output is called income and the surplus or margin of income over costs is called profit. Farmers are then assumed to maximize their satisfactions by maximizing profits, although there are some very important wants such as those for affection, respect and security which cannot be measured objectively in money terms. Most individuals are risk avoiders, which means they will choose less risky activities even though they may not be the most profitable on average over a period of years. Growing traditional crops for home consumption as well as the cash crops may provide protection against various risks; risks of failure of the cash crop, or of falling prices for it, or of rising prices for the staple food crop in local markets. Many farmers would maximize their average profits by growing only cash crops but they maximize their satisfactions by avoiding the risk. Hence in all systems of farming there are various limitations or constraints imposed on the search for maximum profits because of the desire to satisfy other wants not measured in money terms. Nevertheless profits do enable farm families to buy many things which improve diets, health and education, as well as increase the range of choice open to them in material possessions.

There are two broad reasons for expecting farmers to attempt to maximize their satisfactions and profits. Firstly this is the rational way to behave. It would be difficult to explain the behaviour of a man who fails to act in accordance with his preferences; who purposely forgoes benefits or fails to avoid costs. Secondly producers who do not make decisions in this way may not survive in the struggle for existence. Under harsh conditions where farmers are just managing to subsist, those who fail to count their costs and maximize their incomes may starve. Alternatively, commercial farmers who are selling their outputs and buying their inputs in competition with others may be forced out of business if they do not try to maximize their profits. In fact most studies that have been made of African farmers suggest that they do make rational decisions by comparing costs and benefits in relation to their existing technical knowledge and social circumstances; they do attempt to maximize their profits and to minimize their risks and they are unlikely to adopt an innovation if they think that, on average over a period of years, it will add more to costs than to expected benefits.[2]

PHYSICAL PRODUCTION RELATIONSHIPS AND THE ECONOMIC OPTIMUM

The economists' approach to decision making may be illustrated with some results from a detailed study, made by Miss Haswell, of farming in a Gambian village.[3] The author recorded the hours worked by individual farmers on their main cash crop, groundnuts, and the yields they obtained. From these records she was able to establish a relationship between hours worked and output of groundnuts per acre of land, as shown in Table 1.1. This analysis is based on the simplifying assumption that variation in the number of hours worked per acre was the sole cause of variation in the yield of groundnuts. The effects of variation in acreage of land, soil type and the timing of the labour inputs were not specifically taken into account. Nevertheless these results illustrate a fundamental characteristic of physical production relationships, namely 'diminishing returns' to increasing inputs of one resource when applied to a fixed quantity of another.

It will be seen from column II of the table that groundnut yield or total product per acre was increased by increasing labour inputs per acre but the effect of each additional unit of 20 hours of labour was not the same. The amount by which the total product is changed by a single unit change in labour input is called the 'marginal product' per unit of labour and is shown in column

[2] See, for instance, Jones, W. O. (1960) Economic man in Africa, *Food Research Institute Studies*, vol. 1, no. 2, p. 107; Welsch, D. (1965) Response to economic incentive by Abakaliki rice farmers in Eastern Nigeria, *Journal of Farm Economics*, vol. 47, no. 4, p. 900; Dean, E. (1966) *The supply response of African farmers: theory and measurement in Malawi*, Amsterdam, North Holland; Hill, P. (1970) *Studies in rural capitalism in West Africa*, Cambridge University Press, African Studies Series no. 2; Norman, D. W. (1970) *An economic study of three villages in Zaria province: Part 2, Input–output relationships*, Samaru, Nigeria, Institute for Agricultural Research, Samaru Miscellaneous Paper no. 33; and Oni, S. A. (1969) Econometric analysis of supply response among Nigerian cotton growers, *Bulletin of Rural Economics and Sociology*, Ibadan, vol. 4, no. 2, p. 203.

[3] Haswell, M. R. (1953) *Economics of agriculture in a savannah village*, London, H.M.S.O., Colonial Research Study no. 8. Also Haswell, M. R. (1963) *The changing pattern of economic activity in a Gambia village*, London, H.M.S.O., Department of Technical Cooperation, Overseas Research Publication no. 2.

III, rising to a peak at a total labour input of 210 hours per acre then gradually diminishing. The eventual decline in the marginal product per unit of labour does not imply that men do less work per hour if they work longer hours. Even if they work just as hard in each additional hour, the marginal product eventually declines

TABLE 1.1 Diminishing returns to labour: groundnut production, Genieri, the Gambia, 1949

	lb. undecorticated groundnuts		
I Hours worked per acre	II Total product (yield per acre)	III Marginal product per 20 hours	IV Average product per 20 hours
130	250		38·5
		30	
150	280		37·3
		30	
170	310		36·5
		30	
190	340		35·8
		30	
210	370		35·2
		55	
230	425		37·0
		40	
250	465		37·2
		40	
270	505		37·4
		40	
290	545		37·6
		35	
310	580		37·4
		30	
330	610		37·0
		30	
350	640		36·6
		25	
370	665		36·0

Note. Columns I and II show the hours worked per acre and the yield obtained per acre when the inputs were increased by equal units of 20 hours. Column III is obtained from the difference between successive outputs in column II and column IV by dividing the total product by the hours worked, then multiplying by 20.

because it is impossible to increase the output of groundnuts indefinitely by using more and more labour on a fixed area of land. After a point the use of more labour becomes less effective in increasing output. As the marginal product of labour falls, this eventually causes the average product to fall as shown in column IV of the table.

This case of diminishing returns to labour in groundnut production is an example of a general rule, the so-called 'law of diminishing returns' which may be stated as follows. 'If the inputs of one factor are increased by successive equal amounts, the rates of use of all other factors being held constant, then the marginal (and average) product per unit of the variable factor will eventually decline.'

This technical relationship between physical inputs and outputs forms the basis for deciding rationally how many hours to work per acre of groundnut land. At first sight the answer might seem to be 130 hours, the lowest labour input recorded, since this gave the highest average return per unit of labour. Yet most farmers worked more hours per acre than this. The reason was that land suitable for growing groundnuts was scarce and many families only had a small plot available to them. These families could only increase groundnut production by cultivating the fixed area of land more intensively, using more labour per acre and accepting a lower average return per hour of work. However, the majority of them were not prepared to work for less than 1·8 lb. of groundnuts per hour, which amounts to 36 lb. for a unit of 20 hours. A comparison of this cost per unit of labour with the marginal products obtained suggests that the rational choice for these farmers is to work 290 hours per acre. Between 130 hours and 210 hours per acre the marginal product per unit of labour is less than 36 lb. of groundnuts and so too is the average product for inputs of 190 and 210 hours, so it would be irrational to choose a level of labour use within this range. Between 210 hours and 290 hours each additional unit of labour produces more than 36 lb. of groundnuts so the benefits of increasing the hours worked exceed the costs. However, beyond 290 hours each additional unit of labour produces less than 36 lb. of groundnuts which, we are told, was the minimum amount for which the farmers were prepared to work.

Thus profit, measured in pounds of groundnuts, is maximized at a labour input of 290 hours per acre (700 hours per hectare), a fact which is emphasized by calculating the surplus over labour cost per acre at different levels of labour use, as has been done in Table 1.2. Certain other costs, of seed for example, would have to be deducted to arrive at the profit but since we are assuming that all other costs are fixed, the point of maximum surplus is also the point of maximum profit. This point, which is called the economic optimum, has been found theoretically, assuming rational behaviour, but Miss Haswell claims that 'cultivators did in fact stop putting any further labour into cultivation at about this point.

The Gambian male cultivator may not be able to read or write, but he knows the difference between marginal and average product, which is more than can be said for some highly educated accountants.'[4]

TABLE 1.2 The economic optimum: derived from Table 1.1

	lb. undecorticated groundnuts		
I Hours worked per acre	II Total product (yield per acre)	III Labour cost at 1·8 lb. per hour	IV Surplus over labour cost
130	250	234	16
150	280	270	10
170	310	306	4
190	340	342	−2
210	370	378	−8
230	425	414	11
250	465	450	15
270	505	486	19
290	545	522	23
310	580	558	22
330	610	594	16
350	640	630	10
370	665	666	−1

Note. Columns I and II are taken from Table 1.1. Column III is obtained by multiplying column I by 1·8, the cost per hour; column IV is obtained by subtracting column III from column II. The maximum surplus, or 'profit', of 23 lb. groundnuts occurs at a labour input of 290 hours per acre, the point where the marginal return is approximately equal to the cost per unit of input.

Here then we have the basic principle for deciding how much to produce; to expand production so long as the extra income exceeds the extra cost. The economic optimum is reached when the two are equal, although where, as in our example, inputs are varied by relatively large units they may not be exactly equal, but as near as possible. The economic optimum only exists where there are diminishing returns. If marginal products did not decline eventually, farmers could increase their profits indefinitely by using more and more labour per acre. But since the law of diminishing returns is generally found to apply to all productive processes there is usually an economic optimum.

[4] Clark, C. and Haswell, M. R. (1970) *The economics of subsistence agriculture*, 4th ed., London, Macmillan.

This analysis involved only physical units; hours of labour and pounds of groundnuts. Such physical units avoid some of the problems of using units of value expressed in prices, which vary between places and between periods of time. Many of the advanced techniques of farm management being developed today make use of physical measures of inputs and output. However the analysis was only possible because we were given the farmers' subjective valuation of labour in relation to groundnuts. The fact that some of the farmers worked less than 290 hours per acre and some worked more, suggests that not all of them put the same relative value on their labour. In fact, as we have seen, these subjective values may be influenced by the amount of groundnut land the farm family has at its disposal. Other decisions, such as which alternative crops to produce, depend upon the farmers' relative valuations of the various alternatives, for instance how they value a pound of rice in relation to a pound of groundnuts. It would be very difficult, if not impossible, to arrive at a complete picture of all the farmers' values and preferences, and even if this could be achieved, the pattern would vary from one farmer to another. For more general analysis of farm management problems some objective, common unit of value is needed.

SUBSISTENCE FARMING AND NUTRITIONAL VALUES

The main aim of subsistence farmers is to satisfy the food needs of themselves and their families. Although these needs vary according to age, sex, body weight, activity and climate they can be estimated scientifically. For instance the first and most immediate need is for energy to maintain life and bodily activity, a need which is met in varying degree by all foods. The energy value of any food can be measured by laboratory tests in terms of calories per kilogramme, and standard values have already been published for most foods used in Africa.[5] With these standards the total Calories provided by a combination of different foods may be expressed by a single figure as shown for a typical maize-based diet in Table 1.3. Estimates like those shown in the table may be national averages based on food commodity balances or individual family diets based on household consumption surveys. Such studies

[5] The unit more commonly used is the Kilocalorie, written as 'Calorie' with a capital C and equal to 1000 calories. See F.A.O. (1964) *Food composition tables for use in Africa*, Rome.

are used by scientists of F.A.O. and other organizations to test whether particular populations or groups of people are adequately fed.[6] A person who does not get sufficient Calories from his diet is said to be suffering from 'undernutrition'.

Farm management economists and home economists can use data on the Calorie needs of human beings and the Calorie values of different foods to analyse the problems of subsistence farmers and to plan improvements. The decision of what to produce may be resolved by choosing those crops which yield the highest

TABLE 1.3 A maize diet: average daily consumption per person

	Weight in grammes	Calories per kilogramme	Calories
Maize	500	3760	1880
Green vegetables	175	137	24
Pumpkin	175	274	48
Beans	100	3760	376
Bananas	200	640	128
Beef	200	1830	366
Eggs	60	1700	102
Sugar	60	4000	240
Total Calories			3164

After Godman, A. (1962) *Health science for the tropics*, Longmans, London.

Calorie output from the farmers' limited resources. Estimates of the relative output of Calories per hectare are given for some of the common West African food crops in Table 1.4. From these estimates, it is clear that manioc (cassava) is likely to give the highest output of Calories per hectare of land; as this crop also requires relatively little labour, manioc may also give a high output per hour of work, and these two factors explain why so many African families grow the crop, whenever the physical conditions favour it. Unfortunately it contains relatively little protein, so diets based on manioc may not provide sufficient of this body-building material and malnutrition may result.

The decision of how much to produce is determined by the size of the family and its Calorie and other needs, but the use of nutritional values also makes it possible to analyse the effect of

[6] See Dema, I. S. (1965) *Nutrition in relation to agricultural production*, Rome, F.A.O.; F.A.O. (1966) *Food balance sheets, average 1960–2*, Rome; and F.A.O. (1969) *Provisional indicative world plan for agricultural development*, vols. 1–3, Rome.

increasing the size of the labour force, by employing a relative for example. The marginal product of the extra man may be estimated in terms of Calories and compared with the extra Calorie cost of feeding him.

Choice of method of production may be based on Calorie costs also. The actual amount of energy used by people engaged in different activities can be measured scientifically in Calories and measurements have been made for various farm tasks in Nigeria.[7]

TABLE 1.4 Calorie yields per hectare: estimated index for various West African crops

Crop	Index
Millet, sorghum	100
Rice	121
Maize	122
Cocoyams (taro)	125
Sweet potatoes	187
Yams	261
Plantains	290
Manioc (cassava)	421

After Johnston, B. F. (1958) *The staple food economies of western tropical Africa*, Stanford University Press.

Such measurements allow the benefits of new cultivation methods to be assessed in terms of the energy saved. For example, a simple hand-operated planting machine was developed at the University of Ibadan in 1964, as a possible substitute for the traditional method using a planting stick. As part of the evaluation of the machine the operator's energy expenditure per hectare planted was compared with his energy expenditure per hectare using a planting stick. It was found that the use of the machine reduced the Calorie expenditure per hectare to about one-fourth of that for the traditional method.[8]

The use of Calorie measurements alone is not satisfactory for comparing the food values of different crop and livestock products. An adequate balanced diet should also contain sufficient quantities of protein, vitamins and certain essential elements. Farm planning techniques, such as linear programming, can be

[7] Phillips, P. G. (1954) The metabolic cost of common West African agricultural activities, *Journal of Tropical Medicine and Hygiene*, vol. 57, no. 12.

[8] Idufueko, A. O. (1964) *The importance of energy consumption in some farm operations*, University of Ibadan, unpublished dissertation.

used to find the least-cost combination of crop and livestock products which will meet *all* the requirements of a balanced diet, where the requirements are measured in Calories, grammes of utilizable protein, fractions of a gramme of vitamins and so on, and for a purely subsistence farm, cost might be measured by the total hours of work needed.[9]

However, for most purposes of farm analysis and planning it is inconvenient to work with a combination of several different measures of nutritional requirements. Furthermore, many agricultural products, such as cotton, rubber, jute, tea, coffee and tobacco, have no nutritive value. As an alternative, Clark and Haswell suggest that the natural unit for measuring production in a subsistence community is the kilogramme of grain, the output of all other crops and livestock being multiplied by conversion factors to give an 'equivalent' weight of grain.[10] These authors give a table of conversion factors for all the major agricultural products and estimates of subsistence requirements, land rents and transport costs in terms of kilogrammes of grain. The basis of the conversion factors is not clear; the main food crops are apparently converted on the basis of Calorie values with some allowance for content of protein, but non-food crops are converted on the basis of relative average world prices. These grain equivalent values are, therefore, very useful for making international comparisons of agricultural productivity but they are not a good measure of the satisfactions of individual farmers in a particular locality.

SCARCITY AND OPPORTUNITY COST

The most serious disadvantage of using nutritional values is that they take no account of the effect of scarcity. Thus water is perhaps the most essential item of diet, yet in most areas it is so abundant that drinking water is practically valueless. In desert areas it may be valued more highly.

In fact the quantities of most productive resources and the services they can provide over a given time period are limited in total. This means that, given the state of technical knowledge, the output of a particular crop or livestock product can only be increased by withdrawing resources from some other activity. The consequent reduction in the other activity represents a 'cost',

[9] Joy, L. (1966) The economics of food production, *African Affairs*, vol. 65, no. 261, p. 317.
[10] Clark and Haswell, *The economics of subsistence agriculture*.

measuring a loss of satisfaction. This is the essence of the concept of 'opportunity costs' which are costs measured by opportunities forgone. Thus the opportunity cost of 30 hours of labour used in groundnut production might be the value of the maize that could have been produced with this same 30 hours of labour or it might be the value of 30 hours of leisure which are forgone. In fact the strict definition of opportunity cost is 'the *maximum* value that the factor could produce in an alternative use'. Thus, if the farmer values 30 hours of leisure more highly than the maize, or anything else he could produce in that 30 hours, then the value of the leisure represents the opportunity cost of that amount of labour.

This concept of labour cost differs from that based on nutritional needs. In fact nutritional needs set the minimum level of labour return necessary for survival, but in any society with a level of living above this minimum the marginal product and hence the opportunity cost per man is also above this minimum. Nutritional needs underestimate the cost of labour in all but the poorest societies.

Any factor of production which is fully employed has an opportunity cost. Even natural resources such as land and water, which cost nothing to produce, have an opportunity cost once all the available amounts are in use. The opportunity cost of capital may be the value of its alternative product or the value of the consumption forgone in its production. Where no resources are hired or bought and output is limited by the size of the family labour force and the few items of capital they possess, it is clear that these resources have an opportunity cost to the family. On the other hand where resources are hired or bought, the quantity available to a particular farmer may not be limited so long as he can pay the price. Expanding one activity on the farm need not cause a reduction in another activity; groundnut production could be expanded by hiring labour at say 1·8 lb. of groundnuts per hour, rather than of reducing leisure or maize production. However, the price of 1·8 lb. of groundnuts per hour represents the opportunity cost of the hired labour to the community as a whole and is a direct cost to the farmer.

MONEY VALUES

Where, as in most of Africa, there are opportunities for earning money incomes, a case can be made out for basing farm management decisions on the prices ruling in local markets. The transition from pure subsistence farming to commercial farming, where the whole output is sold for money, has been divided into four stages, namely (1) pure subsistence, (2) subsistence plus a small traded surplus, (3) subsistence plus a regular marketed surplus or plus part-time employment in industry or on plantations, and (4) commercial farming.[11] Areas of pure subsistence farming are now very rare in Africa, so in most cases farm management decisions can be based on money values, bringing into the account both the actual cash expenditure on fertilizers and other purchased inputs and the opportunity cost of inputs provided directly by the farm family. By similar reasoning, the opportunity cost of produce which is consumed by the family is the price it could have been sold for.

Market prices, unlike nutritional values, are influenced by the relative scarcity of a commodity. This may be illustrated by considering seasonal variations in food supplies and food prices. In many parts of Africa there are seasonal shortages of food which result from the exhaustion of the product from the previous harvest some weeks or months before the arrival of the earliest crops of the next season. This pre-harvest 'hungry gap' occurs, moreover, at the time of year when farm families are engaged in the heavy physical work of planting and weeding the main crops, so that their output of energy is limited and they may lose weight during the season of cultivation. Data on food consumption in Ilesha, Nigeria, (Table 1.5) shows that Calorie intake is high in November and February, after most crops have been harvested, but has fallen by May when the first rains have started and work on the new season's crops is in progress. For this reason farmers value extra food more highly in May and July, when it is relatively scarce, than in November and February when it is relatively abundant. This is reflected in the fact that farmers store some food at harvest for later consumption and they are prepared to pay higher prices in periods of scarcity. Thus, the average prices for the basic foods in Western Nigeria, yams, maize and gari, were 2·6, 3·5 and 4·4 pence per kilo respectively in November 1960 but by May 1961 they had risen to 6·2, 6·3 and 4·9 pence per kilo.

If farmers are free to sell as much of a particular product as they wish and if they behave rationally then the market price should

[11] See Abercrombie, K. C. (1961) The transition from subsistence to market agriculture in Africa south of the Sahara, *Monthly Bulletin of Agricultural Economics and Statistics*, vol. 10, no. 2, p. 1, reprinted in Whetham, E. H. and Currie, J. I., eds. (1967) *Readings in the applied economics of Africa*, Cambridge University Press. Also see Winter, E. H. (1956) *Bwamba economy*, Kampala, East African Institute of Social Research.

reflect the relative value they place on that product. Consider a farmer who grows maize for his own consumption but also sells some at say 6 pence per kilo. Now if subjectively he values maize at less than 6 pence per kilo he would gain by selling more of his produce; conversely if he values maize at more than 6 pence per kilo he would gain by selling less. Ultimately we would expect him to adjust his sales of maize to the point where he values the last kilo sold at just 6 pence. Incidentally, if he could buy maize more cheaply than he could produce it, then clearly it would be irrational to produce maize at all.

The economic theory of production will be developed in the next three chapters using market prices to evaluate inputs and

TABLE 1.5 Seasonal variations in dietary Calorie intakes: results from four villages in Ilesha, Western Nigeria

	August 1960	November 1960	February 1961	May 1961	July 1961
Total Calorie intake per head per day	1784	2014	2036	1842	1835
Calorie intake as percentage of Calories required	80	91	93	82	82

After Dema, *Nutrition in relation to agricultural production.*

outputs. In order to simplify the analysis we will assume that products are sold and variable inputs are purchased under conditions of 'pure competition', which implies that there are a large number of sellers of each product and a large number of buyers of each input so that no single individual can influence the market price.[12] For purposes of analysis this assumption has the convenient effect that the price per unit of each product and of each variable input can be treated as a constant, regardless of the level of output. Furthermore, at this stage we will assume that producers have perfect foresight so that there is no uncertainty. Later, when we come to discuss risk and uncertainty, this unrealistic assumption will be dropped.

---

[12] The term 'perfect competition' is used more frequently than 'pure competition' in economic literature but the former is avoided here since there are differences of opinion as to its definition.

SUGGESTIONS FOR FURTHER READING

CLARK, C. and HASWELL, M. R. (1970) *The economics of subsistence agriculture*, 4th ed., London, Macmillan.

JONES, W. O. (1960) Economic man in Africa, *Food Research Institute Studies*, vol. 1, no. 2, p. 107.

LIVINGSTONE, I. and ORD, H. W. (1968) *An introduction to economics for East Africa*, London, Heinemann Educational Books.

LÜNING, H. A. (1967) Patterns of choice behaviour on peasant farms in Northern Nigeria, *Netherlands Journal of Agricultural Science*, vol. 15, no. 3, p. 161.

MOSHER, A. T. (1966) *Getting agriculture moving*, New York, Praeger.

ORD, H. W. and LIVINGSTONE, I. (1968) *An introduction to West African economics*, London, Heinemann Educational Books.

SEIDMAN, A. (1969) *An economics textbook for Africa*, London, Methuen.

WHARTON, C. R., ed. (1970) *Subsistence agriculture and economic development*, London, Frank Cass.

WHETHAM, E. H. (1966) Diminishing returns and agriculture in Northern Nigeria, *Journal of Agricultural Economics*, vol. 17, no. 2, p. 151.

WHETHAM, E. H. and CURRIE, J. (1969) *The economics of African countries*, Cambridge University Press.

2 The production function and profit maximization

We will leave the problem of what to produce on one side for the time being and consider the problems of how much to produce and what method of production to use, assuming that the farmer specializes in a single homogeneous product, which we will call grain. The answer to 'how much to produce' has already been illustrated by the example of labour use in groundnut production in the Gambia but now the theory will be developed further.

The technical relationship between resource inputs and the product output is known as a production function. It is determined by local conditions such as soil type and climate and by the techniques used. This means that the production function for grain will vary from place to place and will be altered by any technical change such as the introduction of irrigation or a new variety of seed. In theory, the function should include inputs of all resources such as available soil nutrients, climate, pests and diseases which might influence grain yield. It would be impossible to specify all of them separately so some may be lumped together into a broad category such as land or labour, and others, which are considered unimportant, may be ignored. We should note that inputs are rates of resource use and outputs are rates of production *per unit of time*. The period of time to which the production function applies must be short enough to avoid changes in techniques which would alter the shape of the function, yet long enough to include the whole production process. For annual crops the production function applies to a single cropping season but complications arise with perennial crops such as oil palm for which inputs and outputs are spread over many cropping seasons. This problem will be discussed in a later chapter on capital.

RESPONSE TO A SINGLE VARIABLE INPUT

Initially we assume that inputs of all resources except one are fixed, in order to study the relationship between a single variable input and the product output, sometimes called the factor-product relationship. This may be expressed in terms of total product, average product, marginal product or elasticity of response, as shown for a hypothetical example in Table 2.1. The elasticity of

TABLE 2.1 The factor-product relationship (hypothetical data)

I Units of labour used per year	II Total physical product per year	III Average physical product	IV Marginal physical product	V Elasticity of response
0	0	0		
			7	
1	7	7·0		
			13	1·9
2	20	10·0		
			11	1·1
3	31	10·3		
			6	0·6
4	37	9·3		
			3	0·3
5	40	8·0		
			−1	−0·1
6	39	6·5		
			−4	−0·6
7	35	5·0		

Note. Average product (column III) equals total product (II) divided by units of resource used (I). Marginal product equals increase in total product, i.e. differences between successive figures in column II.

response, shown in the last column of this table, is the relative change in output divided by the relative change in input. It is most conveniently thought of as the percentage change in output resulting from a 1 per cent increase in input and is a pure number, independent of the units in which inputs and outputs are measured.

The total product is always the sum of the marginal products of the units of variable resource used. For example in Table 2.1 the total product from 3 units of labour, 31 bags of grain, is the sum of the marginal products 7 + 13 + 11 bags. This relationship follows from the definition of marginal product as the addition to total product. The results are shown in diagram form in Figure 2.1. Initially, for the first 2 units of labour input, marginal product per unit increases, a situation which may arise because these units of labour are too thinly spread over the fixed quantity of land, and other resources, to grow and harvest the crop properly. Further increases in labour use, beyond 2 units, are associated with diminishing marginal products which eventually are negative for the sixth and seventh units of labour.

(a) *Total product*

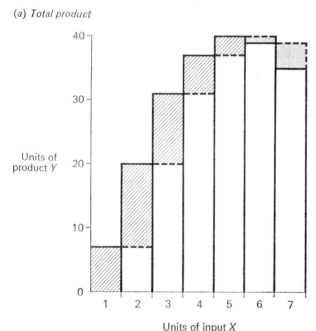

(b) *Marginal product*

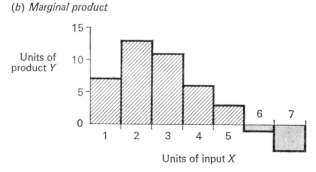

FIGURE 2.1 Diminishing marginal returns. This is a block diagram of the influence of level of input on total product. By convention, in a production function, variable inputs are measured on the horizontal (X) axis and product output on the vertical (Y) axis.

If we assume that labour inputs and grain outputs can be varied continuously, we can illustrate the effect of varying inputs on total, average and marginal products by means of smooth curves as shown in Figure 2.2. The total product curve may be called the 'response curve'. Two important points may be distinguished on the response curve. The first of these occurs at a labour input of

2·6 units where the average product, also known as the productivity, of labour is at a maximum. At this point marginal product is equal to average product and elasticity of response has a value of exactly 1. If land and other fixed inputs were available free of cost

(*a*) *Total product*

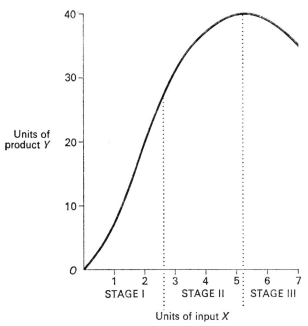

(*b*) *Marginal and average products*

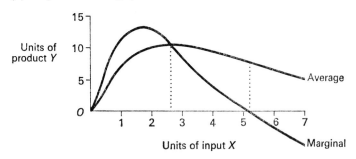

FIGURE 2.2 The response curve. This diagram is simply the result of drawing smooth curves through the block diagram of Figure 2.1. However in (*b*) the average product curve has been added. It is irrational to choose levels of output in Stage I where average product is rising or in Stage III where marginal product is negative.

and the aim was to maximize profit per unit of labour this point would represent the economic optimum.

The second important point occurs at a labour input of 5·2 units where total product is at a maximum. At this point the marginal product per unit of labour is zero and so too is the elasticity of response. This point is sometimes called the technical optimum and it would be the economic optimum also in the unlikely event of labour being free of cost. Because labour normally has a cost the economic optimum normally lies below this level of input.

These two points divide the response curve into three stages, the first and third of which are unlikely to include the economic optimum under any circumstances.[1] These two stages are often eliminated from analysis of farm management decisions as being 'irrational areas of production'. In the first stage of labour inputs below the point of maximum average product, elasticity of response is greater than 1 and the marginal product is greater than the average product. Each extra unit of labour adds more to total product than each of the units already employed so the average product must be increased by employing more labour. In this stage increasing inputs of labour cause increases in both the total product of the given quantities of the fixed inputs and the average product per unit of the variable input. Therefore if it is profitable to produce any output then the farmer can make more profit by using more labour as long as the average product increases. It is therefore irrational to choose a level of input in Stage I.

Similarly it is irrational to choose a level of input in Stage III, beyond the level of maximum total product. Here the marginal product per unit of labour is negative, extra units of labour actually decrease the total product so both the total product of the fixed input and the average product per unit of the variable input are falling at this stage.

Thus we usually find that marginal and average products are positive yet diminishing in practice because most farmers operate in Stage II and avoid the irrational areas. In this second stage, elasticity of response is less than 1 and marginal product is less than average product which therefore decreases as more labour is used. It is possible to discover whether a farmer is operating in the rational area, Stage II, even if we do not know the whole shape of

[1] This need not be true if the farmer could influence prices for inputs and the product by varying his level of output, but this is unlikely to be possible for a small farmer and anyway we are excluding the possibility by assuming pure competition.

the response curve. The only information needed is the average product per unit of the resource in question and the marginal product of just 1 extra unit of input. If the marginal product is positive but less than the average product then marginal and average products must be diminishing.

Over a limited range of input levels, increasing inputs of the variable resource may yield *constant* marginal returns. This would result in a straight line relationship between variable inputs and

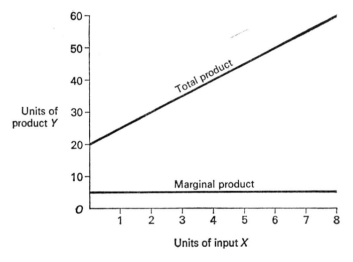

FIGURE 2.3 Constant marginal returns. The 'response curve' is a straight line with a constant slope, i.e. a constant marginal product. However average product may still diminish if, at zero variable input, there is a non-zero product.

total product as shown in Figure 2.3 and might arise where there is an unused surplus of fixed resources. Nevertheless the law of diminishing returns would come into operation eventually once the original surplus of fixed resources was in use.

THE ECONOMIC OPTIMUM

The economic optimum occurs where the value of the marginal product equals the cost per unit of the variable input, which is known as the unit factor cost.[2] Alternatively, this same point may

[2] The economic optimum can only be defined like this if the product is sold and inputs are bought under conditions of pure competition. Where prices vary with level of output, extra income, which is known as the 'marginal value product', may be less than the value of the marginal product (see Chapter 5). Where prices are constant 'marginal value product' is synonymous with 'the value of the marginal product'.

be defined where the marginal product per unit of variable input equals the cost per unit relative to the product value, which in money terms is measured by the inverse price ratio: input price divided by product price. This is illustrated for our hypothetical example by applying a price of £1 per unit of grain and £5 per unit of labour, so the inverse price ratio is £5 divided by £1 which equals 5. In addition the total cost of fixed inputs is assumed to be £10. The resulting product values, costs and profits are given in Table 2.2. For this hypothetical example it appears that the economic optimum occurs at about 4 units of labour input.

TABLE 2.2 Product values and factor costs

I	II Costs			III	IV	V	VI
	£						
Unit of resource input	Fixed costs	Variable factor cost	Total cost	Total value product	Profit	Value of marginal product	Unit factor cost
0	10	0	10	0	−10		
						7	5
1	10	5	15	7	−8		
						13	5
2	10	10	20	20	0		
						11	5
3	10	15	25	31	6		
						6	5
4	10	20	30	37	7		
						3	5
5	10	25	35	40	5		

Note. Derived from Table 2.1. Product price = £1 per bag. Unit factor cost = £5 and fixed cost = £10. Profit is maximized where the value of the marginal product is most nearly equal to unit factor cost, that is at 4 units of resource input.

Below this level the value of the marginal product (column V) is greater than the unit factor cost (column VI) so profit is increased by using more labour. At levels of input above 4 units, profit is decreased because the value of the marginal product is less than the unit factor cost.

The precise point of economic optimum can be found by using the response curve and marginal and average product curves from Figure 2.2, together with lines showing the costs of the inputs in relation to the value of the product as in Figure 2.4. This shows that the economic optimum occurs at 3·7 units of labour input and an output of 36 bags of grain. On the response curve the total cost, made up of a constant fixed cost and an increasing variable cost, is represented by the straight line *CDE*. The difference

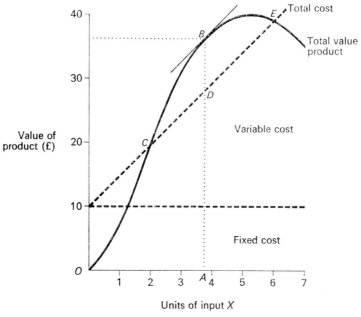

(a) *Total value product and cost*

(b) *Marginal value product and cost*

FIGURE 2.4 The economic optimum level of input. The total cost line has been added to Figure 2.2. Since the product price is £1 per unit, the value product curve is the same as the physical response curve. Profit (the difference between total value product and total cost) is maximized at about 3·7 units of input (point *A*).

between total value product and total cost, *BD*, is profit. It is at a maximum of £7·50 at 3·7 units of labour input where the slope of the total value product curve (which is the value of the marginal product) is equal to the slope of the total cost line (which is the unit factor cost). Note that the slope of the total cost line would be unaltered whatever the level of fixed costs, so the economic optimum is unchanged by changes in the price of the fixed inputs.

On the graph showing the curve of the value of the marginal product, unit factor cost is the same for each extra unit of labour so it is represented by a horizontal straight line. The shaded area minus the cross-hatched area represents the surplus out of which fixed costs must be paid to leave the profit. Since the fixed costs are constant an increase in this area must represent an increase in profit. The area, and hence profit, is maximized at 3·7 units of labour input.

The economic optimum is sometimes called the equilibrium level of inputs, implying that it is a point of balance. Any change in prices or the response curve alters the optimum. For instance if the factor price falls in relation to the product price the economic optimum occurs at a higher level of input and output. In the present example, if the price of labour fell by a half while the price of grain remained unchanged *or* if the price of grain doubled while the price of labour remained unchanged, the economic optimum level of labour input would be increased to 4·4 units, as the reader may confirm by drawing appropriate cost lines in Figure 2.4. Conversely if the factor price rises in relation to the product price the economic optimum occurs at a lower level of input and output. If, in our example, the price of labour increases to £7 per unit or seven times the price per unit of grain, the economic optimum occurs at a labour input of 3·3 units. Should the factor price increase or the product price fall still further it may become unprofitable to produce *any* of that particular commodity.

New techniques of production may alter the production function so that marginal products per unit of labour are increased. If prices are unaltered by the change, the economic optimum level of labour use will be increased. The economic optimum level of output will be increased in two ways, first because the marginal and average product of each unit of labour is increased and second because more units of labour are used. There is a similar effect if more than one input can be varied as discussed in the next section.

MORE THAN ONE VARIABLE INPUT

The illustration used so far, with only one variable resource, is clearly unrealistic. The farmer is able to vary the amounts of many of his inputs. We will therefore consider the effect of varying inputs of two resources. Later in the chapter the conclusions will be generalized to problems involving more than two variable resources.

Again it is helpful to use hypothetical figures to illustrate the relationships involved. The total physical products for different combinations of two resources, land and labour, are given in Table 2.3 in terms of bags of grain. These figures therefore represent the production function relating output to inputs of land *and* labour. Each row in this table represents the response curve for labour, at a given fixed level of the other resource, land. It will be noted that the data used in the top row for 1 unit of land are the same as those given in Table 2.1. At any point in Table 2.3

TABLE 2.3 Production function data: relationship between output in bags of grain and varying combinations of labour (X_1) and land (X_2)

		Units of labour input X_1						
		1	2	3	4	5	6	7
Units of land	1	7	20	31	37	40	39	35
input X_2	2	18	33	45	53	57	57	54
	3	26	42	55	62	65	67	65
	4	31	48	61	68	71	73	72
	5	35	53	66	73	76	78	77
	6	39	58	69	76	79	81	80
	7	40	60	70	77	80	82	81

the marginal product of labour is the difference between two consecutive values in the same row. Each column in the table represents the response curve for land at a given level of labour input. The marginal product per unit of land is the difference between two consecutive values in the same column.

If we assume that inputs of land and labour can be varied continuously, smooth curves can be fitted to the data given in each row or each column. In this case the data represents points on a smooth surface in three dimensions. The response surface for land and labour, based on the hypothetical data from Table 2.3, is given in Figure 2.5. The diagram shows the interlaced single-variable response curves crossing the surface.

We can see from the data given in Table 2.3 and Figure 2.5 that an increase in the second factor, land, causes an upward shift in the response curve for the first factor, labour. This is even more obvious if the separate labour response curves for different levels of land use are drawn on the same two-dimensional diagram, as in Figure 2.6. It is only to be expected that increases in the area of land cultivated will yield increases in total product; indeed

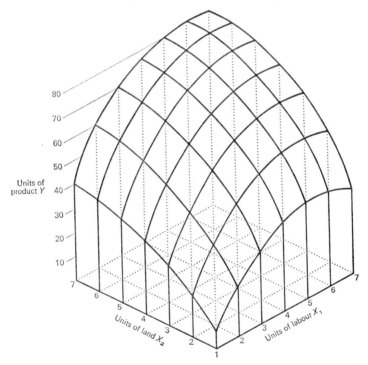

FIGURE 2.5 Response surface for two inputs; with interlaced single-variable response curves. Individual single variable response curves for labour would be represented by vertical slices parallel to the labour (X_1) axis. There are as many different labour response curves as there are possible levels of inputs of land (X_2). If X_2 can be varied continuously, there are an infinite number of response curves for labour.

one might suppose that by increasing land and labour inputs together, diminishing returns would be avoided so that a doubling, for instance, of both land and labour would give double the original output. In fact this is not the case because there are other factors such as capital and managerial inputs which are fixed. Thus the average and marginal products of labour are increased by increasing inputs of land but not in strict proportion; there is an interaction between the inputs.

This interaction between inputs is important since it means that the economic optimum level of labour use depends upon the

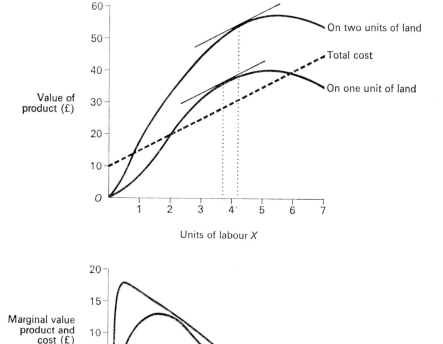

FIGURE 2.6 The influence of land inputs on labour productivity. Total, average and marginal products of labour are increased by using extra land.

quantity of land available. In our example, if we apply the prices of £1 per unit of output and £5 per unit of labour, the economic optimum level of labour input is 3·7 units on 1 unit of land and 4·2 units on 2 units of land. It is therefore misleading to estimate the economic optimum level of use for one variable input alone when others can be varied. Likewise it is misleading to convert labour inputs and product outputs onto a per-hectare or per-acre basis in order to find the economic optimum level of labour use, as was done in the example of groundnut production in the Gambia, discussed in Chapter 1. The overall economic optimum is found when all variable inputs are increased together up to the point where the value of the marginal product equals the unit factor cost for each of them. First it is necessary to investigate the proportions in which variable resources should be combined.

FACTOR-FACTOR RELATIONSHIPS

We will assume in the first instance, not only that two factors are variable, but also that output is fixed at a level below the overall economic optimum. This may arise where the farmer is not concerned with maximizing total profit but wants to produce the necessary food for his family as cheaply as possible. The family food requirements then represent the fixed output level and the problem is to find the least-cost combination of resources.

In order to find the least-cost combination we must consider the alternative combinations of variable resource inputs which will produce the required output of say 40 bags of grain. From Table 2.3 it will be apparent that this level of output could be produced by a combination of 5 units of labour and 1 unit of land *or* by a combination of 1 unit of labour and 7 units of land.

However there are other combinations not specified precisely in this table which would also produce 40 bags of grain; for instance 2 units of labour and between 2 and 3 units of land. These intermediate combinations are given in Table 2.4. Since we have assumed that inputs of labour and land can be varied continuously these data represent points on a smooth curve, joining all combinations of land and labour which give equal levels of output. The curve is therefore called an iso-product curve, the term 'iso' meaning equal. It may also be known as an iso-quantity curve or 'isoquant' for short. In part (b) of Table 2.4 the data represent points on the 50-bag isoquant.

These isoquants are plotted on a graph relating inputs of land to inputs of labour in Figure 2.7. This graph is similar to a contour map used in laying out terraces or in plotting the topography of an area of land. However whereas contour lines on maps represent heights, isoquants are product contours. The isoquant graph is really a plan view of the three-dimensional diagram which is repeated in Figure 2.8 with the isoquants drawn in.

TABLE 2.4 Isoquants and least-cost combinations
Price of labour = £5 per unit
Price of land = £10 per unit
Inverse price ratio 5/10 = 0·5

I	II	III	IV
		Rate of technical substitution:	
Units of	Units of	units of land replaced by	
labour	land	1 unit of labour	Total cost (£)
		$\dfrac{\text{Decrease in } X_2}{\text{Increase in } X_1}$	
X_1	X_2		$£5X_1 + 10X_2$

(a) For 40 bags of grain output

1	7·0		75
		4·2	
2	2·8		38
		1·0	
3	1·8		33
		0·5	
4	1·3		33
		0·3	
5	1·0		35
		−0·1	
6	1·1		41
		−0·2	
7	1·3		48

(b) For 50 bags of grain output

1	—	—	—
2	4·2		52
		1·6	
3	2·6		41
		0·8	
4	1·8		38
		0·3	
5	1·5		40
		−0·2	
6	1·7		47
		−0·2	
7	1·9		54

Returning to the two-dimensional factor-factor relationship represented in Table 2.4 and Figure 2.7, the following characteristics should be noted as they apply to most isoquants for any pairs of factors.

Firstly the slope of the curve shows the quantity of land replaced by one extra unit of labour. It therefore represents the rate of technical substitution.[3] The calculated rate of substitution of labour for land is given in the third column of Table 2.4. It is

---

[3] It is called the rate of technical substitution to distinguish it from the rate of value substitution which depends on the relative prices of the two factors.

obtained simply as the difference between successive values in the second column of the table.

The second point to note is that isoquants generally have a negative slope so, strictly speaking, the slope of the curve is the negative of the rate of technical substitution. This is only to be expected. If we use more of one resource we would hope to use less of the other resource in producing a given level of output. In fact there may be areas where the isoquants slope upwards as is the case for the 40-bag isoquant for labour inputs of over 5 units,

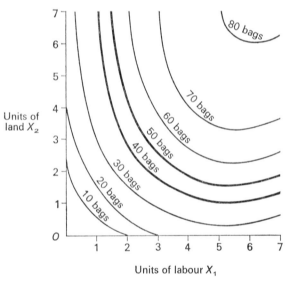

FIGURE 2.7 Isoquants for land and labour in grain production. Isoquants join all possible combinations of land and labour which will produce a given total output. They normally have a negative slope and are convex to the origin.

but these are irrational areas of production. A farmer who uses 7 units of labour and 1·3 units of land to produce 40 bags of grain could obtain just as much output by using less of both factors.

The third point is that isoquants for higher levels of output will normally lie above and to the right of isoquants for lower levels of output. In Figure 2.7, the 50-bag isoquant lies above and to the right of the 40-bag isoquant. This simply means that it will normally require more of either or both resources to produce more output. For similar reasons no isoquants intersect. We would not expect a given combination of land and labour to produce both 40 bags of grain and 50 bags of grain.

The fourth point to notice is that the curves are convex to the origin. This means that the rate of technical substitution diminishes as more of one factor is used to replace the other. In our example each added unit of labour substitutes for, or replaces, less land than the previous unit.

The reason for diminishing rates of substitution is that one factor is rarely a *perfect* substitute for the other. Thus we have

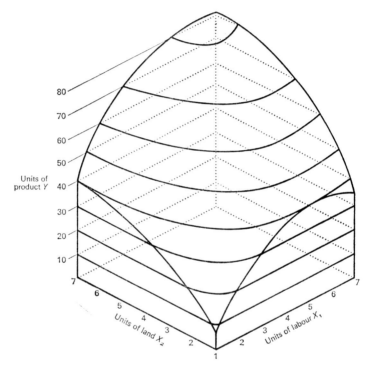

FIGURE 2.8 Product contours on the response surface. The isoquants represent horizontal slices through the surface at specific levels of output.

seen that a given output of grain can be obtained by using relatively little labour and a relatively large amount of land. The same output can be achieved by using more labour and less land but it is difficult to imagine labour substituting entirely for land. In other words it is impossible to produce grain output without using some land. Conversely land can never be substituted entirely for labour. Some labour will always be needed.

Finally, the spacing of the isoquants tells us the effect on total product of increasing inputs of both factors together. If isoquants are equally spaced this means that there are constant returns as the two inputs are increased together. Where, as in the present example for outputs up to 50 bags, the isoquants get closer together, there are increasing returns to increased inputs of both factors. Conversely where isoquants become further apart, as in the present example for outputs over 50 bags, there are decreasing returns to increases in both inputs.

THE LEAST-COST COMBINATION

With information on the rate of substitution between two resources and the respective price of each, the least-cost combination can be estimated. In Table 2.4 the total cost of each combination of land and labour is given in the final column, where labour costs £5 per unit and land costs £10. For 40 bags of grain the least total cost of £33 occurs where 3 or 4 units of labour are used and between 1·3 and 1·8 units of land. The cost is at a minimum because the extra cost of substituting labour for land is less than the saving in land cost. For instance at 2 units of labour the extra cost of 1 more unit is £5, the saving in land cost is £10. It would therefore pay to substitute labour for land. Conversely, if more than 4 units of labour are used, the extra cost of substituting labour for land is greater than the saving in land cost. For instance the cost of the fifth unit of labour is still only £5 but the saving in land cost is only £3. It would no longer be profitable to substitute land for labour.

At this 'least-cost combination of resources' the rate of technical substitution is equal to the inverse price ratio. The inverse price ratio where price of labour is £5 and price of land is £10 is 0·5. The rate of substitution of labour for land is equal to 0·5 for the fourth unit of labour input. As we have seen this level represents the least-cost combination.

For an output of 50 bags of grain the rate of technical substitution equals the inverse price ratio at about 4 units of labour and 1·8 units of land. This therefore represents the least-cost combination.

The argument may be reinforced by reference to the graph of the isoquants. In Figure 2.9 straight equal-cost (or isocost) lines are superimposed on the graph. These lines represent all combinations of the two factors land and labour which could be purchased for a given total cost when the price of land is £10 per unit and the price of labour £5 per unit. For instance the line *AB* represents the isocost for £30 total outlay. This sum could be used to hire 3 units of land and no labour *or* 6 units of labour and no land or any intermediate combination of land and labour along the straight line joining these two points. Similarly the lines *CDE* and *FGJ* represent the isocosts for total outlays of £32·50 and £38 respectively.

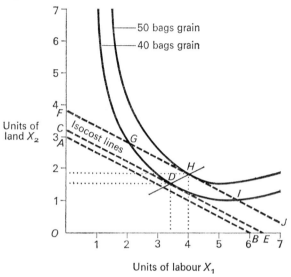

Units of labour X_1

FIGURE 2.9 The least-cost combination and the expansion path. The least-cost combination of labour at £5 per unit and land at £10 per unit for 40 bags of grain is 3·4 units of labour and 1·55 units of land costing £32·50. For 50 bags of grain it is 4 units of labour and 1·8 units of land costing £38.

The slope of the isocost is the number of units of land which can be hired for the price of 1 unit of labour. Its slope is negative since, if total expenditure is kept constant, an increase in the quantity of land hired must be accompanied by a decrease in the quantity of labour. In fact the number of units of land which can be hired for the price of 1 unit of labour depends upon the relative prices of land and labour. More precisely it equals the price per unit of labour divided by the price per unit of land. In the example this is £5 divided by £10 which equals 0·5. Thus the slope of the isocost is the negative inverse price ratio of the two resources. It is independent of the total expenditure. This means that if the relative prices of the two resources remain constant the ratio will remain constant regardless of the quantities used by the individual farmer. For this reason the isocosts are straight lines and parallel to each other.

It should now be clear that the least-cost combination of resources occurs where an isocost is tangent to (i.e. just touches) the isoquant. Thus the least-cost combination for an output of 40 bags of grain occurs at point *D* in the diagram, where total cost is £32·50. At any other combination of resources on this isoquant (e.g. point *G* or point *I*) the total cost is higher (e.g. £38 at *G* or *I*).

The least-cost combination for an output of 50 bags of grain occurs at point *H* where the £38 isocost line just touches the isoquant. At this point the slope of the two lines is the same. The rate of technical substitution equals the inverse price ratio at the least-cost combination of resources. The graphical analysis therefore confirms our earlier conclusion.

For a given set of relative prices we can trace out the least-cost combinations for increasing levels of output. The line joining all least-cost combinations for a given set of prices is known as the 'expansion path'. A small part of it is represented by the line *DH* in Figure 2.9. It traces out the combinations of variable inputs a rational farmer would use as he increases the output of his farm.

MARGINAL PRODUCTS AND COSTS

We have seen that the rate of technical substitution is equal to the inverse ratio of the marginal physical products of the two resources and at the least-cost combination the marginal rate of substitution is also equal to the inverse price ratio. It follows that the ratio of marginal physical products is equal to the price ratio at this point. This in turn means that the marginal product of labour divided by the price of labour is equal to the marginal product of land divided by the price of land. As a result we can now define the least-cost combination in another way; namely the combination where the marginal product per £1 (or other unit of expenditure) is the same for each resource. If we use MPP_1 and MPP_2 to represent the marginal physical products of labour and land respectively and P_1 and P_2 to represent their prices the least-cost combination occurs where $MPP_1/P_1 = MPP_2/P_2$.

That this is indeed the least-cost combination may be demonstrated with the help of the hypothetical data given in Table 2.5. It is assumed that the combination of minimum cost has been found since the ratio of the marginal physical products is equal to the

TABLE 2.5 Marginal products of labour and land: in bags of grain at the least-cost combination

	Labour	Land	Ratio
Price per unit	£5	£10	0·5
Marginal physical product per unit (bags)	8	16	0·5
Marginal physical product per £ (bags)	1·6	1·6	1·0

price ratio and the marginal physical product per £1 expenditure is the same for labour and land. Now suppose that an extra £1-worth of labour is used. This will necessitate using £1-worth less land in order to keep total expenditure constant. Thus more labour will be used on less land. Even if the quantity of land remained constant we would expect diminishing returns to labour, so the gain in product might only be 1 bag of grain instead of 1·6. But the reduction in land use must result in a loss of at least 1·6 bags of grain. If increased use of land results in diminishing marginal returns then *decreased* use of land must result in *increased* marginal returns so £1 *less* spent on land might result in a loss of output greater than 1·6 bags. Hence the loss from spending an extra £1 on labour is greater than the gain. By similar reasoning we would expect an extra £1 spent on land to result in a loss which is greater than the gain. This means that it is impossible to increase the total production of grain by any reallocation of the limited funds between labour and land. But if total production is at a maximum for a given total outlay, cost per unit of outlay must be at a minimum. Thus labour and land must already be combined in the least-cost combination.

Naturally a relatively small change in the price of either resource will lead to a change in the least-cost combination. If the price of labour increases in relation to the price of land, then costs will be minimized by using *less* labour and *more* land. This may be checked by adding isocosts to Figure 2.9 for land at £10 per unit and labour at say £7·50 per unit. The converse will be true, if the relative price of land rises.

New techniques of production may alter the shape of the isoquants so that a given level of output can be obtained with less of one or other or both the resources. In the present context an innovation which reduces the inputs of both labour and land needed to yield a given output of grain would be classified as neutral. One, such as the introduction of new seed and fertilizers, which tends to reduce the amount of land needed rather more than the amount of labour, would be called a land-saving innovation. Mechanization on the other hand is usually labour-saving.

PERFECT SUBSTITUTES AND COMPLEMENTS

In a few cases, factors of production may be perfect substitutes for each other. For instance this could be true of compost and a complete chemical fertilizer. For some farm operations, possibly weeding, female labour may substitute entirely for male labour. In such cases it almost invariably pays to use only one of the two resources, namely the cheaper. If compost is cheaper, per unit of nutrients, than the compound fertilizer then only compost should be used. If female labour is cheaper than male labour then only female labour should be used for weeding. This is true even if women cannot work as fast as men, provided the cost per unit of work is lower.

This is illustrated in Figure 2.10 which shows the isoquants (as solid lines) for given levels of weeding output, assuming that men can work half as fast again as women. This means that $1\frac{1}{2}$ units of female labour are necessary to do the work of 1 unit of male labour. Similarly 3 units of female labour substitute for 2 units of male labour. Since male labour and female labour are considered to be perfect substitutes, the isoquants are straight lines.

Now if we assume that men are paid twice the hourly wages of women we can see that women should be employed for weeding rather than men. The relative cost of weeding by women is $1.5 \times 1 = 1.5$ units and the relative cost of weeding by men is $1 \times 2 = 2$ units therefore the total cost is higher if men are used.

In Figure 2.10 the isocosts are shown by the broken lines, with the slope $-P_1/P_2 = -2$. In this diagram isocosts are never tangents to the isoquants; rates of technical substitution are never equal to price ratios. The least-cost method of production is therefore found at the end of an isoquant where it touches the lowest isocost line. For the lower isoquant in the diagram, *AB*, the least-cost method is found at point *A*, using $1\frac{1}{2}$ units of female labour only. All other points on *AB* lie on higher isocost lines. Point *B* represents the most costly method. If the cost of male labour was less than $1\frac{1}{2}$ times that of female labour then the least-cost method would be to use only male labour. If the inverse price ratio was exactly 1·5, that is the same as the productivity ratio, then it would not matter whether men or women were used.

At the other extreme from perfect substitutes are perfect complements among factors of production which means that one is useless without the other. An increase in only one of the complementary inputs without a corresponding increase in the other will yield no extra product. Alternatively a reduction in the amount

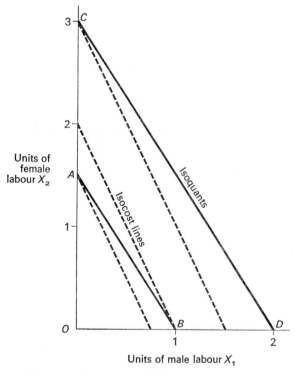

FIGURE 2.10 Isoquants for perfect substitutes. For perfect substitutes the isoquants are straight lines. The least-cost method then involves only one of the two resources. In this hypothetical example female labour is the cheaper.

available of one of the inputs will render some units of the other factor unproductive. For example it might be argued that labour is complementary to hoes; although men can work without hoes it is difficult to see how hoes could be used productively without men to wield them. Thus if a farmer buys more hoes without providing additional labourers there will be no increase in production. Conversely if the farmer loses one of his labourers, the man's hoe is no longer productive.

In such a case there is a single least-cost combination of resources regardless of relative prices. These fixed factor proportions will probably hold at all levels of output. Over a wide range of relative prices and levels of output the ratio of one hoe to one man is the cheapest. It is then best to think of the two resources together as a single resource, in deciding the economic optimum level of use.

MORE THAN TWO VARIABLE INPUTS

The rule for determining the least-cost combination is easily extended to more than two variable resources. However, the production function cannot then be drawn in two dimensions.

The least-cost combination of three resources is found when

$$\frac{MPP_1}{P_1} = \frac{MPP_2}{P_2} = \frac{MPP_3}{P_3}$$

where *MPP* represents marginal product and *P* represents price per unit of each resource. This equation can be extended to cover any number of resources by adding extra terms.

Expressed in words, the least-cost combination of any number of resources occurs where the marginal product per £1 (or other unit of expenditure) is the same for each resource. This then is the solution to the problem of what method of production to use.

However, it is only necessary to consider the combination of resources if output differs from the economic optimum. If *every* variable resource is used up to the point where marginal value product equals unit factor cost, then the marginal product per £1 spent will automatically be the same for every variable resource. But if *every* resource is varied, an economic optimum may not exist, as a result of constant or increasing returns to scale. When all inputs are expanded together we talk of an increase in scale. For example, if the amount of each resource used is doubled, then there is a doubling in scale; a 5 per cent increase in scale would mean that the amounts of all resource inputs were increased by 5 per cent.

RETURNS TO SCALE

Strictly speaking, the law of diminishing returns only applies where the inputs of *one* factor are increased, the quantities of all other factors being held constant. When inputs of two or more variable factors are increased together, the marginal physical products are unlikely to diminish as rapidly as when a single resource is increased alone. For instance the marginal product of labour will diminish less rapidly if each extra unit of labour has additional land to work with. However, even when two or more factors are increased together, returns are likely to diminish eventually so long as at least one other factor is fixed. But when there is an increase in scale, all inputs are varied together and none are fixed. The law of diminishing returns no longer applies.

If two inputs are increased together by 1 per cent we expect output to increase by the sum of their elasticities of response. For instance if the elasticity for land is 0·2 and for labour 0·6 then if inputs of both land and labour are increased by 1 per cent, output will increase by 0·2 per cent plus 0·6 per cent, which is 0·8 per cent. Similarly the effect of increasing *all* inputs by 1 per cent (that is, increasing scale by 1 per cent) is obtained as the sum of all the elasticities of response. This sum of elasticities of response is sometimes known as the scale coefficient. Now if the scale coefficient is equal to 1, it means that a 1 per cent increase in scale leads to a 1 per cent increase in output. There are constant returns to scale. If the sum of elasticities is less than 1 there are decreasing returns to scale and if the sum of elasticities is greater than 1 there are increasing returns to scale.

It is tempting to assume that production functions will necessarily exhibit constant returns to scale. If all inputs are doubled one might expect output to be doubled also. This would simply represent a duplication of the original unit.

There is an important consequence of constant or increasing returns to scale; namely that if prices remain constant there is *no* economic optimum level of production. If all resources are varied together, marginal returns will not diminish for any resource, so provided the farmer can go on obtaining resources and selling his products at the same price, there is nothing to prevent him from increasing his profits by expanding his output indefinitely.

SUGGESTIONS FOR FURTHER READING

DILLON, J. L. (1968) *The analysis of response in crop and livestock production*, Oxford, Pergamon Press.

DOLL, J. P., RHODES, V. J. and WEST, J. G. (1968) *Economics of agricultural production, markets and policy*, Homewood, Illinois, Irwin.

HEADY, E. O. (1952) *Economics of agricultural production and resource use*, Englewood Cliffs, Prentice-Hall.

VINCENT, W. H., ed. (1962) *Economics and management in agriculture*, Englewood Cliffs, Prentice-Hall.

Mathematical appendix to Chapter 2

SINGLE VARIABLE INPUT

Let X = input of variable resource and
   Y = output of product.

Now since for every value of X there is a corresponding value
of Y we may assume that the two are related in some systematic
manner. In fact we assume that Y depends, at least in part, on X.
In mathematical symbols we may write $Y = f(X)$ which means Y
is a function of X and does not represent some number 'X'
multiplied by another number 'f'. As, in this case, we are concerned
with a production relationship, we call it a production function.

Such a production function is represented by the data given in
Table 2.1 where the figures in the first column are values of X
and those in the second column the corresponding values of Y.
It may be possible to express the function as a specific algebraic
relationship such as $Y = 3X^{0.4}$ or $Y = 6 + 10X - 2X^2$ from
which we can calculate the corresponding table of values, but
this will not always be the case.

From the production function we can derive three quantities of
interest:

(1) *the average physical product* of X, written APP_X;
(2) *the marginal physical product* of X, written MPP_X;
(3) *the elasticity of response* with respect to X, written E.

The first two of these measures are physical quantities; the
third, E, is a pure number.

The relationships between total, average and marginal products are illustrated in Figure 2.11 which is simply a reproduction of the response curve shown in Figure 2.2.

Average physical product is defined as

$$APP_X = \frac{Y}{X}.$$

Thus the average product at point A on the graph is the product
associated with 1 unit of input i.e. $7/1 = 7$ bags of grain. This is the
slope of the straight line OA which connects point A with the
origin. The average product at point B is exactly the same since
the slope of the straight line OB which connects point B with the
origin is the same i.e. $39.9/5.7 = 7$ bags of grain.

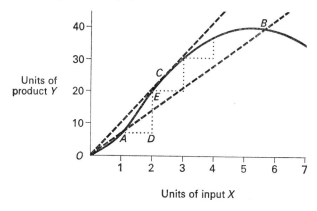

FIGURE 2.11 Factor-product relationships. Marginal product equals the slope of the total product curve. Between A and E this equals ED/AD but $AD = 1$ therefore slope equals ED, the vertical distance. As we move up the curve the vertical distance diminishes, i.e. marginal product diminishes. At point B marginal product is negative. Average product equals Y/X, i.e. the slope of a straight line through the origin. Average products at A and B are the same. Average product is at a maximum at point C.

The maximum average product occurs where the line through the origin just touches (i.e. is a tangent to) the total product curve. This is obvious from the graph where C is the point of tangency, the line OC having the highest slope of any average product line. Maximum average product is obtained in this case at about 2·6 units of input.

If we use the symbol Δ to mean a change, so that ΔY is a change in Y and ΔX is a change in X, then marginal physical product is defined as

$$MPP_X = \frac{\Delta Y}{\Delta X}.$$

This is the slope of the total product curve itself. For instance the marginal product of the second unit of input is the slope of the line AE which equals $ED/AD = 13/1 = 13$ bags of grain.[4]

Reference to Figure 2.11 should make it clear that at point C, where average product is at a maximum, the two slopes are equal, so $\Delta Y/\Delta X = Y/X$, i.e. $MPP_X = APP_X$ at this point. At levels of

[4] Readers with a knowledge of differential calculus will note that if ΔX is measured in very small units, in fact as it approaches zero, then $\Delta Y/\Delta X = dY/dX$ which is the first derivative of Y with respect to X. If the production function may be expressed as a specific algebraic relationship then marginal products may be derived directly by differentiation.

input below 2·6 units in this graph, where marginal and average returns are increasing, the slope of the total product curve is steeper than the slope of the average product line which means that $\Delta Y/\Delta X$ is greater than Y/X, i.e. MPP_X is greater than APP_X. Conversely at more than 2·6 units of input, marginal and average returns are diminishing; $\Delta Y/\Delta X$ is less than Y/X and MPP_X is less than APP_X.

The total product Y is at a maximum of 40 bags, at $X = 5\cdot3$ units of input in Figure 2.11. At this point the total product curve is horizontal, so $\Delta Y/\Delta X$, i.e. $MPP_X = 0$. At levels of input above 5 units Y decreases as X increases so $\Delta Y/\Delta X$ (or MPP_X) is negative.

Elasticity of response with respect to X is measured as the relative change in Y divided by the relative change in X, thus

$$E = \frac{\Delta Y}{Y} \Big/ \frac{\Delta X}{X} = \frac{\Delta Y}{\Delta X} \Big/ \frac{Y}{X} \text{ i.e. } \frac{MPP_X}{APP_X}.$$

Since MPP_X and APP_X are measured in similar units E is a pure number. Since where there are increasing returns MPP_X is greater than APP_X, E must be greater than 1. Conversely where there are diminishing returns MPP_X is less than APP_X so E is less than 1.

THE ECONOMIC OPTIMUM

Now let $P_X =$ price per unit of X (unit factor cost),
      $P_Y =$ price per unit of Y, and
  $MVP_X =$ value of the marginal product of X in the production of Y.

The economic optimum may now be expressed in several ways.

1. At the economic optimum extra cost equals extra return:

$$MVP_X = P_X \tag{1}$$

i.e. value of the marginal product equals unit factor cost.

2. Dividing both sides by P_X gives

$$\frac{MVP_X}{P_X} = 1 \tag{2}$$

i.e. value of the marginal product in £ per £1 spent is unity.

3. $MVP_X = MPP_X \cdot P_Y$ by definition.
Substituting this in equation (1) gives

$$MPP_X \cdot P_Y = P_X.$$

Dividing both sides by P_Y gives

$$MPP_X = \frac{P_X}{P_Y} \qquad (3)$$

i.e. marginal physical product equals the inverse (factor-product) price ratio.

4. $MPP_X = \dfrac{\Delta Y}{\Delta X}$ by definition.

Substituting this in equation (3) gives

$$\frac{\Delta Y}{\Delta X} = \frac{P_X}{P_Y} \qquad (4)$$

i.e. marginal rate of transformation of X into Y must equal the inverse price ratio.

5. Multiplying both sides of equation (4) by

$$\frac{P_Y \cdot \Delta X}{\Delta Y}$$

gives

$$P_Y = \frac{\Delta X}{\Delta Y} \cdot P_X; \qquad (5)$$

but if the individual producer receives a constant price for his product, price of product equals marginal revenue per unit of product: $P_Y = MR_Y$, also $\Delta X / \Delta Y = $ quantity of X required to produce 1 extra unit of Y. Hence if the individual producer pays a constant price P_X for the variable resource, then the quantity of X required to produce 1 extra unit of Y times the price of X is the marginal cost of 1 extra unit of Y, thus

$$\frac{\Delta X}{\Delta Y} \cdot P_X = MC_Y$$

Therefore equation (5) means

$$MR_Y = MC_Y$$

i.e. marginal revenue equals marginal cost.

TWO VARIABLE INPUTS

Now let X_1 = quantity of variable resource (1) used,
X_2 = quantity of variable resource (2) used, and
Y = total physical product obtained.

The production function may now be written as

$$Y = f(X_1, X_2).$$

As with a single input there might be an algebraic formula which would describe the function, such as: $Y = 3X_1{}^{0.4} . X_2{}^{0.6}$, but this need not be the case. In fact the product Y depends upon more than two resources so we should write $Y = f(X_1, X_2, X_3, \ldots, X_n)$. However we are now assuming that all resources except the first two are fixed so we write $Y = f(X_1, X_2/X_3, \ldots, X_n)$. The average physical product of each variable input is now defined as

$$APP_1 = \frac{Y}{X_1},$$

$$APP_2 = \frac{Y}{X_2}.$$

Similarly the marginal physical products are defined as

$$MPP_1 = \frac{\Delta Y}{\Delta X_1},$$

$$MPP_2 = \frac{\Delta Y}{\Delta X_2}.$$

Elasticities of response are defined as

$$E_1 = \frac{\Delta Y}{Y} \Big/ \frac{\Delta X_1}{X_1},$$

$$E_2 = \frac{\Delta Y}{Y} \Big/ \frac{\Delta X_2}{X_2}.$$

The rate of technical substitution of resource (1) for resource (2) is defined as

$$RTS_{12} = -\frac{\Delta X_2}{\Delta X_1}.$$

It will be noted that for a given level of output X_2 will increase as X_1 decreases.

For a small increase in inputs of resource (1), ΔX_1, the change in output would be $\Delta X_1 . MPP_1$. Likewise for a small decrease in inputs of resource (2), ΔX_2, the change in output would be $-\Delta X_2 . MPP_2$. But if total output remains constant these two changes must be equal, so that they cancel out. Thus we may write

$$\Delta X_1 . MPP_1 = -\Delta X_2 . MPP_2.$$

Dividing both sides by $\Delta X_1 . MPP_2$ gives

$$\frac{MPP_1}{MPP_2} = -\frac{\Delta X_2}{\Delta X_1} = RTS_{12}.$$

That is, the rate of technical substitution equals the ratio of the marginal products.

THE LEAST-COST COMBINATION

Now let $P_1 =$ unit factor cost of X_1,
                $P_2 =$ unit factor cost of X_2,
                $P_Y =$ price per unit of Y,
        $MVP_1 =$ value of the marginal product of X_1 in
                         producing Y,
and $MVP_2 =$ value of the marginal product of X_2 in
                         producing Y.

The least-cost combination may now be expressed in several ways.

1. The least-cost combination occurs where extra cost (negative gain) of substituting resource (2) for resource (1) just equals the extra saving (positive gain):

$$-\Delta X_2 . P_2 = \Delta X_1 . P_1. \tag{6}$$

2. Dividing both sides by $P_2 . \Delta X_1$ gives

$$-\frac{\Delta X_2}{\Delta X_1} = \frac{P_1}{P_2}.$$

But $$-\frac{\Delta X_2}{\Delta X_1} = RTS_{12}.$$

Therefore $$RTS_{12} = \frac{P_1}{P_2} \tag{7}$$

i.e. rate of technical substitution of resource (1) for resource (2) equals the inverse price ratio.

3. *But*
$$RTS_{12} = \frac{MPP_1}{MPP_2}.$$

Hence by substituting in equation (7) this gives
$$\frac{MPP_1}{MPP_2} = \frac{P_1}{P_2}.$$

Multiplying both sides by $\dfrac{MPP_2}{P_1}$ gives
$$\frac{MPP_1}{P_1} = \frac{MPP_2}{P_2} \tag{8}$$

i.e. marginal physical product per £1 spent is the same for each resource.

4. But both sides may be multiplied by the same quantity, P_Y, without changing the equality. By definition
$$MPP_X \,.\, P_Y = MVP_X.$$

Hence we can write
$$\frac{MVP_1}{P_1} = \frac{MVP_2}{P_2} \tag{9}$$

i.e. value of the marginal product per £1 spent is the same for each resource.

5. Equation (9) is consistent with the rule for finding the economic optimum level of resource use which states
$$MVP_X = P_X \text{ or } \frac{MVP_X}{P_X} = 1 \text{ (from equation (2))}.$$

Hence $\quad \dfrac{MVP_1}{P_1} = \dfrac{MVP_2}{P_2} = 1$ (compare with equation (9)).

RETURNS TO SCALE

Let us assume that there are three, and only three, variable resources[5] involved in the production of output Y, then
$$Y = f(X_1, X_2, X_3).$$

The elasticities of response with respect to X_1, X_2 and X_3 are given by:

Note the argument is readily extended to any number of resources.

$$E_1 = \frac{\Delta_1 Y}{Y} \bigg/ \frac{\Delta X_1}{X_1},$$

$$E_2 = \frac{\Delta_2 Y}{Y} \bigg/ \frac{\Delta X_2}{X_2},$$

$$E_3 = \frac{\Delta_3 Y}{Y} \bigg/ \frac{\Delta X_3}{X_3}.$$

$$(10)$$

The subscripts after the Δ signs are necessary to denote that the marginal products differ for the three resources.

Now if all three resource inputs are increased at the same rate K then

$$\frac{\Delta X_1}{X_1} = \frac{\Delta X_2}{X_2} = \frac{\Delta X_3}{X_3} = K.$$

The three equations (10) may now be rewritten as

$$E_1 = \frac{\Delta_1 Y}{Y} \bigg/ K,$$

$$E_2 = \frac{\Delta_2 Y}{Y} \bigg/ K,$$

$$E_3 = \frac{\Delta_3 Y}{Y} \bigg/ K.$$

Hence

$$K \cdot E_1 = \frac{\Delta_1 Y}{Y},$$

$$K \cdot E_2 = \frac{\Delta_2 Y}{Y},$$

$$K \cdot E_3 = \frac{\Delta_3 Y}{Y}.$$

$$(11)$$

Now the overall rate of change in Y is the sum of these three terms (11):

$$\frac{\Delta Y}{Y} = \frac{\Delta_1 Y}{Y} + \frac{\Delta_2 Y}{Y} + \frac{\Delta_3 Y}{Y}$$

$$= K \cdot E_1 + K \cdot E_2 + K \cdot E_3 = K(E_1 + E_2 + E_3).$$

But $e = E_1 + E_2 + E_3$ by definition, therefore

$$\frac{\Delta Y}{Y} = K \cdot e$$

$$(12)$$

i.e. rate of increase in output equals rate of increase of inputs multiplied by the scale coefficient.

(*a*) *Constant returns to scale.* Now if $e = 1$, equation (12) becomes

$$\frac{\Delta Y}{Y} = K = \frac{\Delta X_1}{X_1} \tag{13}$$

i.e. rate of increase in output equals rate of increase of inputs. The same is true for resources (2) and (3).

Equation (13) implies that

$$\Delta Y = K \cdot Y \text{ and } \Delta X_1 = K \cdot X_1.$$

Hence the new average product

$$\frac{Y + K \cdot Y}{X_1 + K \cdot X_1} = \frac{Y(1 + K)}{X_1(1 + K)} = \frac{Y}{X_1}$$

i.e. average product remains constant and there is generally no economic optimum if prices are also constant.

(*b*) *Increasing returns to scale.* Now if *e* is greater than 1 equation (12) implies that

$$\Delta Y \text{ is greater than } K \cdot Y.$$

Hence the new average product

$$\frac{Y + \Delta \cdot Y}{X_1 + K \cdot X_1} \text{ is greater than } \frac{Y + K \cdot Y}{X_1 + K \cdot X_1} = \frac{Y}{X_1}$$

i.e. average product is increasing and there is generally no economic optimum if prices are constant.

(*c*) *Decreasing returns to scale.* If *e* is less than 1 equation (12) implies that

$$\Delta Y \text{ is less than } K \cdot Y.$$

Hence the new average product

$$\frac{Y + \Delta Y}{X_1 + K \cdot X_1} \text{ is less than } \frac{Y + K \cdot Y}{X_1 + K \cdot X_1} = \frac{Y}{X_1}$$

i.e. average product is diminishing and an economic optimum will normally exist even with constant prices.

3 Cost analysis and optimum size

Returns to scale are only obtained when *all* inputs are increased at the same rate, that is when no resources and hence no costs are fixed. A situation where no resources are fixed occurs so rarely in agriculture that true increases in scale are not often found. When some resources are fixed and others are variable or where inputs of *all* factors are increased or decreased, but at different rates, then the changes in the production relationships are not ones of pure scale since the factor proportions vary. Any increase in the total quantity of resources used can be described as an increase in farm size, whether factor proportions vary or not, so an increase in scale is a special case of an increase in farm size.

We are then faced with the problem of measuring farm size. Most frequently it is spoken of in terms of hectares of land; a 3-hectare farm is said to be larger than a 2-hectare farm. This measure is unsatisfactory because land is only one of several factors that determine the size of the farm business. As we have seen from the study of groundnut producers in the Gambia, output per acre (or per hectare) could be varied a great deal according to the amount of labour used. The intensity of land use, by which we mean the quantity of labour and other resources used per hectare of land, can be varied. Thus intensive methods of production using irrigation and fertilizers on 2 hectares of land may require more labour and produce more output than extensive (i.e. un-intensive) methods of production such as shifting cultivation on 3 hectares or an even larger area of land.

To some extent the problem of varying intensity may be over-come by measuring farm size in terms of labour inputs, but these will depend upon the extent of mechanization as well as the level of output. This difficulty may be avoided by using standardized labour requirement estimates to arrive at a 'standard man-day' measure of size. Farm size is then measured by labour require-ments assuming that all farms use the same methods of production, an obvious unreality where large and small farms are adapted to different methods. Nevertheless farm size is measured in this way by the British Ministry of Agriculture. Ideally all factors of production should be taken into account, but it is difficult to combine different inputs into a single measure of size. Where there is a single uniform product, the best solution is to use total quantity produced to measure size, although this may be influenced by chance weather variations. This measure will be used for the remainder of this chapter.

There are certain advantages associated with increasing size of business, such as increasing scope for mechanization. These advantages are often misleadingly, called 'economies of scale'. At the same time there may be some disadvantages of large farms, which are called 'dis-economies of scale'. Some of these advantages and disadvantages may be felt by other farmers in the district, where, for instance, one farm or estate grows enough cotton to justify the establishment of a ginnery which can be used by others or where one farmer keeps so many cattle that the grassland becomes overgrazed and erosion results. These are known as external economies and dis-economies.

It should be noted that small farmers may gain some of the advantages of larger farms by cooperation in marketing their products, in sharing machinery or even in joining up to form a communal farm. However the dis-economies that limit the size of individual farms may also affect cooperatives. A large farmer may be able to obtain advantages for himself in the prices he receives for his product or pays for his inputs. Merchants who deal with farmers are involved in greater costs and usually make higher charges per ton of seeds, fertilizer or other chemicals if these materials are sold in small lots; the accounting expenses are increased and so too are transport costs through having to deliver over a wider area. There may also be wastage. Banks generally find it more convenient and less costly to make relatively large loans. They may therefore charge higher interest rates for small loans.

Similarly excessively high costs are incurred in selling produce in very small lots. It costs less per ton to send a truckload of groundnuts from Kano to Lagos by railway than to send them in separate consignments, each of which must be handled, documented and delivered individually. Furthermore with larger quantities of produce it is possible to sort into grades and hence in many cases to obtain higher prices. Thus as size is increased there may be scope for reduction in the costs of buying and selling. Marketing cooperatives may obtain some of the advantages of buying and selling in large amounts.

However, at this stage of the analysis we assume that farmers buy and sell under conditions of pure competition so that no individual can influence prices on his own. This leaves us with certain internal economies and dis-economies which are caused by either resource indivisibility or resource fixity.

RESOURCE INDIVISIBILITY

A great many farm inputs cannot be subdivided into small units; they are chunky or indivisible. For instance a tractor cannot be subdivided, neither can a man or a cow. A farmer cannot own three-quarters of a tractor or employ $1\frac{1}{2}$ men, other than by cooperative sharing. Of course, if the services of these resources can be hired by the hour as and when desired, then inputs can be varied continuously, but frequently this is not possible and the choice is between employing or not employing a whole tractor, man or cow.

In some cases indivisible resources are available in several different sizes; for instance there are small two- and three-wheeled tractors as well as four-wheeled tractors and crawlers. However, two- and three-wheeled tractors cannot do all the jobs that four-wheeled tractors do and invariably cost more to buy and operate, per unit of power. Similarly buildings such as grain stores can be constructed in various sizes but up to a point the larger they are, the cheaper they are per cubic metre of space. This is, at least in part, because the surface area of the walls does not increase as fast as the volume enclosed. For instance doubling the length of the side walls of a rectangular building increases the capacity of the building four times.

Another way in which indivisible resources become more effective on large farms is by specialization and division of labour. Even on a small family farm there is some division of labour between different members of the family. For instance the men may do the clearing and cultivation and the women the weeding and livestock husbandry. However, on a large farm this specialization can be taken further so that each worker can become more expert at his particular task. Whereas a small tree-crop producer must carry out all the work himself, on a large plantation there may be specialist nurserymen, specialist pruners, specialist tractor drivers and mechanics and so on. Similarly on a large farm special purpose machinery such as harvesting machinery may be justified more easily than on a small farm.

Where resource inputs are indivisible then the production function will be discontinuous as shown in Figure 3.1. The shape is very different from the smooth response curve depicted in Figure 2.2, though that represented a smoothing out of the discontinuous function shown in Figure 2.1. In that case the units

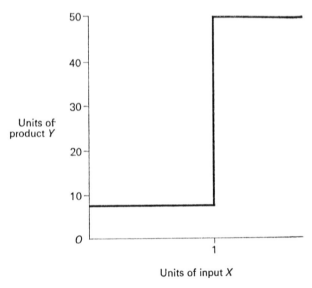

FIGURE 3.1 A discontinuous production function. Where the input cannot be subdivided it may not be possible to adjust production to the economic optimum. Marginal value product will then differ from unit factor cost.

of input were small in relation to the whole range of inputs under consideration so the smooth curve was a reasonable approximation. Now we are considering situations where the indivisible units of input are large in relation to the farmers' range of choice which makes it highly unlikely that marginal value product can be equated with unit factor cost.

Consider for instance the introduction of oxen and a plough into an area hitherto cultivated in scattered plots by hand tools. Let us assume that the total annual cost of employing the oxen and plough is £50 and, of course, this cost cannot be subdivided. Now, clearly, if the extra returns are expected to be less than £50 it would be irrational to introduce the plough team. Since it would be a strange coincidence if the extra return was expected to be exactly £50 the farmer would generally need to expect greater returns than this to be persuaded to use oxen. In order to provide grazing and fodder crops for his animals as well as earning the extra income he may need to increase his farm area from perhaps 1 or 3 hectares under hand cultivation up to say 6 hectares.

Having acquired one team of oxen the farmer might consider using a second team but now he would require a total additional return, over hand cultivation, of more than £100 to justify employing two plough teams. He might then be farming 20 hectares or more. This means that the economic optimum level of use of the indivisible input of a plough team is fixed over a relatively wide range of land inputs (from 6 up to 20 hectares) and product outputs. The inputs and hence the costs of the plough team are fixed, not in any absolute sense but simply because it does not pay to vary them.

Now if one team of oxen could plough up to 20 hectares per season then clearly if it only ploughs 6 hectares it is under-employed; the oxen are not working anywhere near their full capacity. This means that their work capacity is being wasted and their cost is high per unit of output. On a larger farm the average fixed cost per unit of output would be lower. Work capacity of men and machines may be wasted on farms which are too small to keep them fully employed. Thus there are certain advantages obtained from having a large enough output to justify employing an ox team or a hired worker or a particular machine. There are further advantages to be gained from having a large enough output to keep the oxen, the man or the machine fully employed.

RESOURCE FIXITY

Other inputs may be fixed because of past decisions. This means that the resources are already committed and the costs incurred are now unavoidable. For instance once land has been cleared or tree crops planted the necessary resources have already been used so the costs of these inputs are fixed. Contractual agreements, such as regular hiring agreements, are another way in which resources become committed in advance. Sometimes a part of the cost may be avoided, if things go wrong and the farmer wants to cut his losses. We then say the resource has a salvage value, but this is likely to be less than the original cost of the input. Cleared land or growing tree crops may be sold or hired to others in some circumstances. Contracts may be broken if compensation is paid. Thus in some cases only a part of the original cost is unavoidable. Of course, these resources are only fixed over a limited period of time. Eventually the productivity of cleared lands or tree crops declines and new land or trees are needed to replace them. Costs can then be avoided by not replacing them. In fact in this sense fixity is a relative term. Thus if we are planning what to produce in the following year the acreage of tree crops is best considered as fixed. If we are planning ten years ahead then the acreage of tree crops is clearly variable. On the other hand, when planning how to allocate labour over the next few days, then the acreage of all crops must be considered as fixed.

For this reason some economics textbooks suggest that *all* resources are variable in the long run; that is provided a long enough period elapses. However, the indivisible resources discussed above do not depend upon the time period. A farmer might go out and buy a team of oxen, a plough or a tractor immediately. These resources are fixed because it does not pay to vary them. The marginal value product of the last unit in use exceeds its cost but the marginal value product of one more unit would be less than its cost.

Furthermore some resources are fixed in a more absolute sense. The total supply of land is virtually fixed and once it is all under cultivation the area available may set a limit on the scale of farming possible. On the individual farm the managerial ability of the farmer is ultimately a limiting factor. A particular cause of resource fixity in Africa is the lack of markets in productive resources. Today there are well established markets both for export crops and some local food crops in many parts of Africa, but the scope for buying and selling (or hiring) productive resources is frequently limited. This is true of land, labour and capital. In many areas it is not possible to buy or hire additional land, and this sometimes means that a farmer who wishes to expand cannot acquire more land. Then again it may be impossible to hire labour. Where this is so the labour force is restricted to that available from the family and is therefore fixed. Cooperation and communal work sharing generally do not enable an individual farmer to increase his total labour inputs. Where there is no system of moneylending or rural credit, the farmer is forced to rely upon his personal savings or efforts to increase his capital resources.

Finally we must recognize that each farmer is himself a 'fixed input' for his own business since he provides the management. Under pure competition with no risks and uncertainties the farmer may not be faced with the task of day-to-day decision making but even under these rather artificial conditions he would still have the task of supervision of his work force. Where this is limited to the family supervision is fairly easy, but as a farmer increases the size of his farm then he frequently increases the area which he must supervise and the number of people he must organize. If plots are scattered all over the village fields, detailed supervision becomes more difficult, the quality of the work falls, output per man may fall while costs per unit of output rise. The farmer perhaps spends more time in marketing, and has therefore less time to give to the close observation of his farm upon which efficiency so much depends. Farmers vary in their managerial ability but for any one farmer this ability sets a limit on the extent to which he can expand the size of his farm. Managers may be hired in some instances but then if two or more managers are hired, there is the task of organizing and supervising the managers. Eventually a hierarchy of managers is needed and management costs grow much faster than the output of the business. This has occurred on many government-owned plantations and schemes.

The fixity of some resources means that it is impossible to increase all resources to scale. This means that there are certain advantages of increasing size because as the fixed costs are spread over a greater total output, the cost per unit of that output must fall. On the other hand, because these resources are fixed, the law of diminishing returns will apply and the marginal physical products of the variable resources will eventually fall. This is the main disadvantage of increasing size which may set a limit on the growth of farms.

COST RELATIONSHIPS

The influence of fixed inputs on the economic optimum level of production can be better understood by studying the 'cost function' relating cost of production to the level of product output, rather than by studying the production function, which relates output to levels of inputs. The cost function can be calculated from a knowledge of the production function and the prices of all inputs.

However, total costs and hence average costs include fixed costs. If the fixed costs really do not change over a considerable range of output, then average fixed cost per unit of output must diminish as output is increased. Thus by spreading the fixed costs (or over-head costs as they are sometimes known in this context) over a greater total product, the average cost is reduced. Fixed costs have no influence on the marginal cost which depends solely on the variable inputs.

These relationships are illustrated in Table 3.1 and Figure 3.2

TABLE 3.1 Cost function data

Total product (bags of grain)	Total fixed cost ($£$)	Total variable cost ($£$)	£ per bag of grain			Marginal cost
			Average fixed cost	Average variable cost	Average total cost	
5	10	3	2·00	0·60	2·60	0·55
10	10	5·50	1·00	0·55	1·55	0·50
15	10	8	0·67	0·53	1·20	0·43
20	10	10	0·50	0·50	1·00	0·40
25	10	12	0·40	0·48	0·88	0·45
30	10	14·50	0·33	0·48	0·81	0·55
35	10	18	0·29	0·52	0·80	0·90
40	10	25	0·25	0·62	0·87	2·00

Note. The marginal cost given in the last column is the extra cost of producing one more bag of grain. Since the fixed cost is unaltered, the marginal cost is the extra variable cost.

which are derived directly from Table 2.2. The following charac-teristics should be noted.

1. The relationship between marginal and average costs is precisely the same as that between marginal and average products. Thus, if the marginal cost is below average cost then the latter must fall. If the marginal cost is above average cost then the latter must rise. Average cost is at a minimum where it is equal to mar-ginal cost.

2. At low levels of output marginal costs are below average costs which are therefore **diminishing**. This is partly the result of increasing marginal and average products of the variable factor and partly the result of spreading the fixed cost.

3. Eventually marginal and average costs rise as a result of diminishing marginal returns to the variable factor.

In this example we assumed that only one input, namely labour, was variable but by working in terms of costs per unit of output it is possible to take account of any number of variable inputs for which we know the relative prices. It is necessary to assume that the problem of variable input combinations has been solved and that total product is increased by increasing farm size along the expansion path. If all inputs may be increased to scale and there are constant returns to scale, then marginal and average costs per

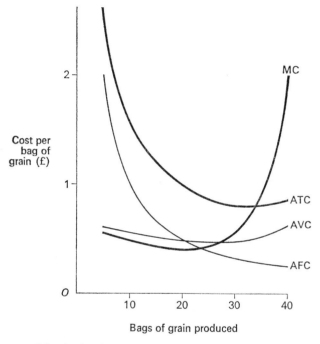

FIGURE 3.2 Marginal and average cost curves. MC = marginal cost, AVC = average variable cost, ATC = average total cost, AFC = average fixed cost. Note the rapid decline in AFC which causes an initial fall in ATC.

unit of product will be equal and constant. On a graph the marginal and average cost curves would in fact appear as a single horizontal straight line.[1] As we have seen, if some resources are fixed then average total cost may fall initially but eventually will rise because of the effect of diminishing marginal returns. Hence, where many resources are varied, so long as some others are fixed, the average and marginal cost graph may not differ significantly from Figure 3.2. The average cost curve would however be a

---

[1] This would be an average variable cost line as no costs would be fixed.

rather wider U-shape, since we would expect marginal and average costs to increase less rapidly than where only one resource is varied.

THE ECONOMIC OPTIMUM

As before, the economic optimum is reached when the extra cost of expanding production is just equal to the extra benefit. In this case we are considering the marginal cost of one more unit of output. Therefore, at the economic optimum this should just equal the marginal revenue from one more unit of output. But, provided that the farmer is not operating on a large enough scale to influence prices, marginal revenue may be taken as a constant. In fact it is the price received per unit of output.

For the costs of grain production given in Table 3.1, if grain sells for £1 per bag, then for levels of production up to 10 bags the fixed cost per bag exceeds the value of the product. In fact a farmer would lose money if he produced anything less than 20 bags of grain, since the average total cost would exceed the price per bag. This serves to emphasize the importance of spreading the fixed cost over a larger quantity of output. The economic optimum occurs at an output of about 35 bags of grain, where the marginal cost is approximately equal to the marginal revenue or price per bag. At this level, total profit is £7 (35 × £1 = £35 minus 35 × £0·80 = £28). The reader may check that different levels of output listed in the table would yield less profit than this.

This optimum level may be shown on the graph of marginal and average product curves by adding the marginal revenue line as in Figure 3.3. The horizontal line *AB* represents the marginal revenue or price of £1 per bag. It crosses the marginal cost line, so the two values are equal, at a total output of 36 bags of grain. The total revenue then is £36, which is measured by the area *ABCO*. Total cost equals average cost per bag times the number of bags produced so it is measured by the area *DECO*. Profit, which is the difference, is the area *ABED*. This area is at a maximum where marginal cost equals marginal revenue. Clearly, if there were constant returns to scale and the marginal cost line was horizontal it would never equal marginal revenue, unless the two happened to be equal at all levels of product. In either case there would be no single economic optimum.

The economic optimum we have now defined differs from that discussed for a single variable input in the previous chapter. That

merely established the level of a single variable input that produced the maximum profit, but as we saw the optimum for one
variable input depends upon the levels of other variable inputs.
Now in considering costs we are assuming that all variable inputs
are increased together so the optimum level of output is an overall
optimum size of business.

The minimum product price below which it is no longer profit-

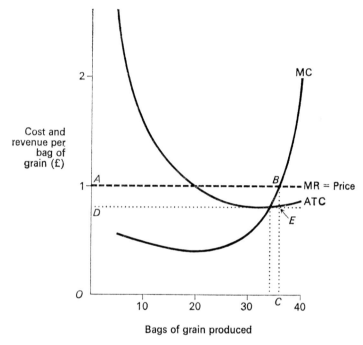

FIGURE 3.3 The economic optimum level of output. The economic optimum
occurs where MC = MR. For a price per unit of output of £1 this occurs at
36 units of output.

able to continue production is the price at which marginal cost
equals average cost. This occurs when 34 bags of grain are produced
at an average and a marginal cost of £0·80 per bag. If the price
received per bag is also £0·80 then clearly no profit would be
made. It might appear that production is not justified at this
level if *no* profit is made. However, if the cost of *every* factor of
production, including capital and management, is included in
total and average costs, then any profit above this would, in a
sense, represent a surplus.

If the product price falls below the minimum average total cost then the total revenue is insufficient to pay the cost of all the productive resources used. That being so the farmer would avoid loss by ceasing production of that enterprise and using his resources in their best alternative use. Of course all resources should be costed at their opportunity cost which may differ from the original purchase price. Thus, many economics textbooks wrongly suggest that the price of the product can fall below the minimum average *total cost*, in fact, down as far as the minimum average *variable cost*, before the producer should stop production entirely. This conclusion is wrong because it assumes that the fixed resources have no opportunity cost. Many fixed resources *do* have an opportunity cost either in terms of what they could earn in alternative enterprises on the farm or in terms of their salvage value. If the opportunity cost is not covered in its current use then clearly its current use is not the most profitable.

However, even for a single enterprise there may be a series of average and marginal cost curves, if there are several alternative techniques of production. Thus hand cultivation will have its own set of fixed costs and its own average and marginal cost curves. Ox-draft cultivation will have a different average and marginal cost curve even for the same product. Costs of tractor cultivation will be different again. The graph of these cost curves might appear as in Figure 3.4. The hypothetical average cost curve for hand labour is marked LL', that for ox-draft is marked DD' and that for tractor TT'. We see then that if the farmer expects to produce and sell X_2 tons of grain, the least-cost method of production will be that based on ox-draft. In fact below output X_1, average cost per unit is lowest for hand labour, between X_1 and X_3 it is lowest for ox-power and above X_3 it is lowest for tractors.

This leads to the interesting conclusion that it may not pay to expand up to the economic optimum with a particular technology. Before the economic optimum is reached a new method may be introduced with lower average costs. In our example the economic optimum level of grain production using hand labour is X_2 tons *but*, as we have seen, before this level of production is reached it is cheaper and therefore more profitable to use oxen.

An actual study of the relative average costs of different methods of cultivation was made in Uganda in the 1950s.[2] Comparison was made of the average cost per acre of cultivating coffee farms of various sizes by hand labour, by a small tractor and disc harrow

[2] Joy, L., ed. (1960) *Symposium on mechanical cultivation in Uganda*, Kampala, Argus.

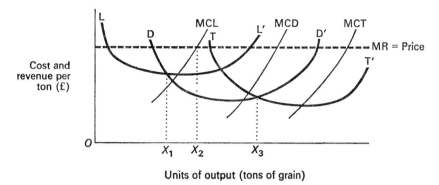

FIGURE 3.4 Farm size and method of production. Up to the output level X_1 hand labour is the cheapest method of production. From X_1 to X_3 ox-draft is the cheapest method and from X_3 upwards tractor-power has the lowest average cost.

costing 12,000s. and by a standard tractor and disc harrow costing 17,400s. The results are illustrated graphically in Figure 3.5. If we ignore the possible influence of different cultivation methods on yields, acres may be used as a measure of output. Hence this graph is comparable with Figure 3.4. It differs, however, in that it was assumed that hand labour can be hired as and when required so average labour cost per acre does not vary with acreage

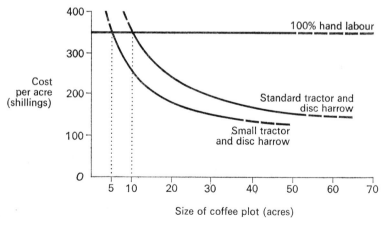

FIGURE 3.5 Comparative costs of hand labour and machinery for varying coffee plot sizes. A small tractor might pay its way on coffee plots larger than 6 acres. On about 8–9 acres the saving could amount to 100s. per acre per year. At about 28 acres the saving could be 200s. per year.

cultivated. As a result the small tractor appears cheaper than hand labour on plots above 6 acres in size and cheaper than the standard tractor on all acreages up to the limit of its working capacity (calculated at a maximum of 80 acres).

This assessment applied to coffee cultivation in Uganda some years ago and may not be directly applicable under other conditions. Furthermore the authors took pains to point out that it was only tentative. Nevertheless it does serve to illustrate how the least-cost method of production varies with level of output.

SUGGESTIONS FOR FURTHER READING

DILLON, J. L. (1968) *The analysis of response in crop and livestock production,* Oxford, Pergamon Press.

DOLL, J. P., RHODES, V. J. and WEST, J. G. (1968) *Economics of agricultural production, markets and policy,* Homewood, Illinois, Irwin.

HEADY, E. O. (1952) *Economics of agricultural production and resource use,* Englewood Cliffs, Prentice-Hall.

HEADY, E. O., JOHNSON, G. L. and HARDIN, L. S., eds. (1956) *Resource productivity, returns to scale and farm size,* Iowa State University Press.

VINCENT, W. H., ed. (1962) *Economics and management in agriculture,* Englewood Cliffs, Prentice-Hall.

4 Choice of enterprise

THE PROBLEM OF WHAT TO PRODUCE

Most farmers are faced with a range of alternative crops they could grow and livestock they could keep. The different sections of the farm, each devoted to the production of one kind of crop or live-stock, are known as enterprises. A problem arises in choosing between different enterprises. This must in turn involve the decision whether to concentrate on the production of one or two enterprises or whether to produce many. In other words there is a choice to be made between a specialized and a diversified system of farming. The economic principle for choosing what to produce is the principle of 'comparative advantage', which states simply that each unit of resources should be used where it will earn the greatest return.

In theory, if a farmer has access to unlimited resources this principle need not apply. In such circumstances enterprises would be independent of each other; an increase in one would neither aid nor hinder another. Opportunity costs of resources would be identical with their purchase prices, so profits would be maximized by expanding each enterprise up to the economic optimum. However, this is an unrealistic view as there will almost invariably be some resources which are fixed in supply. Now enterprises will compete with each other for *some* of these fixed resources so that expansion of one enterprise must be accompanied by a reduction in another. As a result the number of enterprises which it is profitable to combine in a farming system will be limited, in some cases to one enterprise only.

It should be noted that enterprises do not necessarily compete for *all* fixed resources. Even where the total amount of land available is fixed, swamp rice would probably not compete with sorghum for land because the two crops require different types of land. Early maize may not compete with cotton for land since the two crops are grown at different seasons of the year. Thus certain fixed resources are specific to individual enterprises, while others may be allocated between enterprises.

A SINGLE RESOURCE AND TWO PRODUCTS

The problem of finding the most profitable enterprise combination is best illustrated in the first instance by a hypothetical example in which just two alternatives are considered, grain production and groundnut production. These enterprises compete for the labour resources of the farmer and his family which amount to 7 units in total. It is assumed that other resources specific to each enterprise are fixed so that as labour inputs are increased, the marginal returns diminish.[1] Hypothetical total and marginal products for labour in each of the two enterprises are given in Table 4.1.

TABLE 4.1 Labour productivity in two alternative enterprises

| | Bags | | | |
| | | | | |
Units of labour used	Total physical product of grain	Marginal physical product of grain	Total physical product of groundnuts	Marginal physical product of groundnuts
1	7	7	6	6
2	20	13	12	6
3	31	11	18	6
4	37	6	24	6
5	40	3	29	5
6	39	−1	30	1
7	35	−4	30	0

In order to find the optimum combination of the two enterprises, prices are applied to give the marginal value products for labour. Grain is priced at £1 per bag and groundnuts at £3·50 per bag to give the results shown in Table 4.2. If the wage for labour in alternative occupations is £5 per unit, this may be taken as the unit factor cost. The economic optimum level of labour use in grain production then occurs at about 4 units, where the marginal value product is £6. *But* since only 7 units of labour are available in total, then the use of 4 units in grain production would leave

---

[1] Although the total labour resources are *fixed* at 7 units, the quantity used on either enterprise is *variable* between the limits of 0 and 7 units. Hence the law of diminishing returns will apply.

only 3 units for use in groundnut production. Now labour use in groundnut production is below the economic optimum which occurs at about 5 units of input. Hence the two enterprises compete for the scarce labour resources and labour has an opportunity cost within the farm.

TABLE 4.2 Marginal value products in two alternative enterprises

Units of labour	£ Marginal value product of grain at £1 per bag	Marginal value product of groundnuts at £3·50 per bag
1	7	21
2	13	21
3	11	21
4	6	21
	3	17·50
	−1	3·50
7	−4	nil
Total value product	35	105
Average value product	5	15

Starting from the point where 3 units of labour are used in groundnut production, the marginal value product of one more unit of labour is £21. The opportunity cost, represented by the loss in value of grain caused by removing the fourth unit of labour from the grain enterprise, is £6. This transfer will add to profits since extra benefits exceed extra costs. In fact the principle of comparative advantage tells us to use this unit of labour in groundnut production, since this is where it will add most to returns.

If another unit of labour is transferred from grain production to groundnut production, the marginal value product is £17·50 and the opportunity cost £11. Thus transfer of labour is still profitable since the gain is greater than the cost. It should be noted that because the total amount of labour is limited, its opportunity cost at £11 is now well above the wage rate of £5. Now 5 units of labour are used in groundnut production and only 2 units in grain production. The transfer of yet another unit of labour into groundnut production will not increase profits, since the gain would be only £3·50 and the opportunity cost £13. There is no longer a comparative advantage in groundnut production!

Thus the most profitable combination of the two enterprises occurs where about 5 units of labour are used in groundnut production and about 2 units in grain production.

The same conclusion is reached if we consider a unit-by-unit increase in labour input. To maximize profits each unit of labour should be used where it will add most to returns. Therefore the first unit should be used in groundnut production where it will earn £21. Similarly the second, third, fourth and fifth units should all be used in groundnut production. However the sixth unit of labour would earn only £3·50 in groundnut production, whereas if used in grain production it will earn £7. Thus this unit of labour should be used in grain production. The seventh unit of labour should also be used in grain production, where it will earn £13, as opposed to the £3·50 it would earn if used to produce groundnuts.

If the labour inputs in either enterprise could be varied continuously then the most profitable allocation would occur where marginal value product is equal in each enterprise. Since the steps by which labour is varied are quite large in our example, the marginal value products are not equal, but they are as near as possible. The principle involved in allocating resources between alternative uses is therefore sometimes known as the equal marginal returns principle.

It should be noted that using *all* the resource in the enterprise yielding the highest average value product will not maximize profits. In the above example, average value product is higher in groundnut production at £15 per unit of labour. If all 7 units of labour are used in this enterprise the total return would be £105. If on the other hand 5 units are used in groundnut production and 2 units in grain production, as suggested above, the total return is £121·50.

THE PRODUCTION POSSIBILITY CURVE

The same problem may be approached in another way by considering the alternative combinations of grain and groundnuts which may be produced with a fixed amount of 7 units of labour. Naturally if all 7 units are used in grain production no labour is available for groundnut production. If 6 units of labour are used in grain production then 1 unit is available for groundnut production and so on. Thus the combinations of groundnuts and grain which can be produced with different allocations of the 7 units of labour may be read off from Table 4.1. The first four columns of Table 4.3 are obtained directly from Table 4.1.

TABLE 4.3 Choice of enterprise

Price of groundnuts = £3·50 per bag
Price of grain = £1 per bag
Inverse price ratio = 3·50/1 = 3·5

I	II	III	IV	V	VI
Units of labour used in groundnut production	Units of labour used in grain production	Total physical product groundnuts (bags)	Total physical product grain (bags)	Rate of product transformation: bags of grain per bag of groundnuts	Total revenue (£)
X	$7 - X$	Y_2	Y_1	$\dfrac{\text{Decrease in } Y_1}{\text{Increase in } Y_2}$	$\begin{array}{c} £1 Y_1 \\ + 3·50 Y_2 \end{array}$
0	7	0	35		35
				−0·7	
1	6	6	39		60
				−0·2	
2	5	12	40		82
				0·5	
3	4	18	37		100
				1·0	
4	3	24	31		115
				2·2	
5	2	29	20		121·50
				13·0	
6	1	30	7		112
				∞	
7	0	30	0		105

The quantities of groundnuts and grain given in columns III and IV which can be produced with 7 units of labour represent points on the 'production possibility curve'. If labour inputs on each enterprise can be varied continuously then the production possibility curve will appear as a smooth curve as shown on a graph in Figure 4.1. This curve represents the maximum production possibilities for a given resource level. It is therefore sometimes known as a production possibility frontier, boundary or envelope.

If less labour is available or if labour is used inefficiently, the combination of groundnuts and grain produced will lie inside the production possibility curve such as at point *A*. Conversely if more than 7 units of labour are available the new production possibility curve will lie *outside* the curve for 7 units of labour. Certain characteristics of the production possibility curve should be noted as they apply to most such curves for any pair of products.

Firstly, as with isoquants, the slope of the curve is a measure of the rate at which one commodity can replace another, only in this case we are considering products rather than factors. As resources are transferred from grain production to groundnut production, in effect grain is 'transformed' into groundnuts. The rate at which this occurs is called the 'rate of product transformation'.

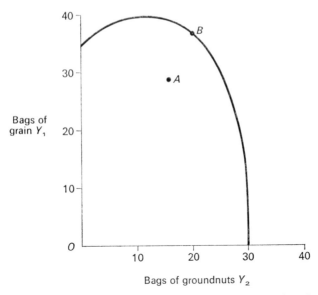

FIGURE 4.1 The production possibility curve. This is the frontier of possible combinations of grain and groundnuts which can be produced with the available resources. Any point inside the curve such as *A* indicates that resources are not fully employed in the best known way. A larger quantity of groundnuts and more grain could be produced at point *B*.

Secondly, it should be noted that over much of the curve the slope is negative; more groundnuts can only be produced at the expense of less grain. Thus the enterprises are said to be 'competitive' since they compete for scarce resources. The rate of product transformation is therefore defined as the negative of the slope of the production possibility curve. Unlike isoquants however, the production possibility curve is concave to the origin, meaning that it bulges outwards. The rate of transformation of one product into another increases as more resources are transferred. As we transfer labour from grain production to groundnut production, the sacrifices of grain output per additional bag of groundnuts become larger and larger. This is a result of diminishing marginal returns. As more labour is put into groundnut production the marginal returns diminish. At the same time as *less* labour is used in grain production the marginal returns increase since we are moving back *down* the response curve. This means that the quantity of grain given up for each unit of labour transferred to groundnut production must increase as more labour is transferred. Conversely, for similar reasons, if we transfer labour from groundnut production to grain production the sacrifices of groundnut output per additional bag of grain become larger and larger.

In particular it should be noted that the first unit of labour used in grain production does not cause any decline in groundnut production and the rate of transformation is infinite. This is because the marginal product of the seventh unit of labour used in groundnut production is zero. Here an enterprise can be expanded without causing a reduction in the other enterprise so they are said to be *supplementary enterprises*. Clearly supplementary enterprises do not compete for resources.

Another interesting section of the curve occurs between zero and 12 bags of groundnut production. Over this section as groundnut production is increased so too is grain production. Whenever the production of one enterprise increases as the other is increased, these are called *complementary enterprises* for as long as this relationship holds. In this example the enterprises are complementary because the marginal returns in grain production are negative. As less labour is used in grain production the total product actually increases. This is not a good example because, as we have seen, it would be irrational to use labour in grain production at levels where marginal returns are negative, whether groundnuts are produced or not. Other, more realistic cases of complementary enterprises will be discussed later.

THE ECONOMIC CHOICE OF ENTERPRISE

With information on the relative prices of the two products the most profitable combination of enterprises can be determined. In Table 4.3 the total revenue from each combination of groundnuts and grain is given in the final column, where grain is priced at £1 per bag and groundnuts at £3·50 per bag. As we have already seen the maximum total revenue of £121·50 occurs where 5 units of labour are used in groundnut production and 2 units are used in grain production, that is where 29 bags of groundnuts and 20 bags of grain are produced. Costs remain the same since 7 units of labour are used in all the situations given, so the total profit is maximized at the point where total revenue is maximized.

This maximum profit point may be defined in terms of the rate of product transformation in a manner which is similar to the definition of the least-cost combination of resources. Thus the maximum profit combination of enterprises occurs where the rate of product transformation of Y_1 for Y_2 equals the inverse price ratio. In the example given in Table 4.3 the price of grain is assumed to be £1 per bag and the price of groundnuts £3·50 per bag. The inverse price ratio is therefore $3·50/1 = 3·5$. The rate of product transformation is nearest to this value at 2·2 where 5 units of labour are used in groundnut production and 2 units are used in grain production. As we have seen, this is the maximum profit allocation of labour between these enterprises.

Finally this profit-maximizing position is illustrated graphically in Figure 4.2 which represents the production possibility curve from Figure 4.1 together with iso-revenue lines. These lines join all combinations of groundnuts and grain which yield the same total revenue. The lowest iso-revenue line shown in the diagram represents all combinations of groundnuts and grain which will yield a total revenue of £35. This can be obtained from £35/£3·50 = 10 bags of groundnuts or £35/£1 = 35 bags of grain or any combination of the two products along this line. The next iso-revenue line shown in the diagram is the £70 iso-revenue line. The third one is for a total of £121·50. All iso-revenue lines are parallel provided that the relative prices of the products do not change.

The slope of the iso-revenue lines is negative because for total revenue to remain constant, increased revenue from one product must be associated with decreased revenue from the other. In fact the number of bags of grain which must be sold to earn the same revenue as a bag of groundnuts depends upon the relative prices of grain and groundnuts. More precisely it equals the price per bag of groundnuts divided by the price per bag of grain. In the example this is £3·50 divided by £1 which equals 3·5. Thus the slope of the iso-revenue line is the negative inverse price ratio of the two products.

The maximum revenue that can possibly be obtained is £121·50 since the iso-revenue line for this sum of money just touches (is tangent to) the production possibility boundary. All other feasible product combinations on, or within, the boundary must lie on lower-valued iso-revenue lines. Thus the optimum product combination is found at point Z where 20 bags of grain and 29 bags of groundnuts are produced.

The maximum-revenue combination of enterprises would be altered by any change in the price ratio of the two products. If the relative price of groundnuts fell it would pay to produce less groundnuts and more grain. It should be noted, however, that for the production possibility curve given, it will never be profitable to

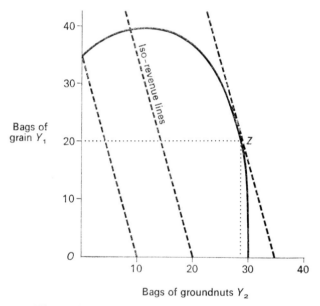

FIGURE 4.2 The optimum choice of enterprises. The maximum possible revenue is obtained at point Z where the iso-revenue line is tangent to the production possibility curve.

concentrate entirely on one enterprise, that is to specialize. Even if groundnut prices per bag rose to say 100 times the price of grain per bag it would still pay to produce at least 7 bags of grain. Conversely if grain prices rose to a very high level in relation to groundnut price it would still pay to produce 10 bags of groundnuts. This may be checked by adding iso-revenue lines to Figure 4.2, for various alternative price ratios. This advantage in favour of diversification, or producing both enterprises, is due to supplementary and complementary relationships between them.

Any technical change which altered labour productivity in one enterprise more than in the other would alter the optimum combination of products. For instance if labour productivity in grain production increased more than labour productivity in groundnut production, proportionately more grain would be produced at the optimum.

MORE THAN TWO ENTERPRISES AND MORE THAN ONE RESOURCE

In this hypothetical example only two enterprises were considered but the conclusions are readily extended to any number of alternative products. The maximum-profit combination of grain and groundnuts was found where the marginal value product of labour was the same in each enterprise. Hence, if there is a third enterprise the maximum profit will occur where labour use is so adjusted that the marginal value product is equal in each of the three enterprises. Similarly the equi-marginal return principle can be applied to any number of enterprises. This is the general solution to the product-product problem.

However, we have so far been considering a rather unrealistic case, where only the labour resource may be allocated between enterprises. It has been assumed that other resource inputs are specific to individual enterprises and are fixed. This fixity of other resources led to diminishing returns as more labour was applied to either of the two enterprises and hence caused the concave production possibility curve. This in turn meant that the maximum profit system involved both enterprises.

In practice most resources can be allocated to any of several enterprises. Not only labour is allocated in this way, but also land, managerial inputs and many items of capital such as tools and machinery. In fact there may be no fixed resources which are specific to individual enterprises. In that case, as more resources are concentrated in one enterprise, returns to scale may be obtained. If there are decreasing returns to scale in each enterprise, it is still likely that the production possibility curves will be concave and a diversified system will be the most profitable. Each resource should then be allocated so that its marginal value product is the same in each enterprise. If, however, there are constant or increasing returns to scale in a particular enterprise, specialization on that enterprise might yield the highest return for a given quantity of resources.

SPECIALIZATION

If, for example, there are increasing returns to scale in groundnut production, then any transfer of resources into other enterprises will cause a greater relative decline in the quantity of groundnuts produced. If one-tenth of the resources used in groundnut production are transferred to grain production the total quantity of groundnuts produced would fall by more than one-tenth.

In such a case, the production possibility curve could be a

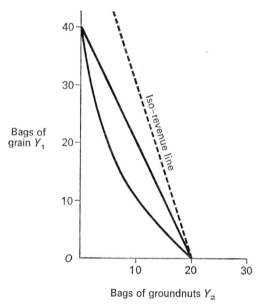

FIGURE 4.3 Constant or decreasing rates of product transformation. The maximum possible revenue is obtained by specialization in a single product; in this case groundnuts.

straight line or even a curve which is convex to the origin. Both these possibilities are illustrated in Figure 4.3 for the two products groundnuts and grain. The straight line is based on the assumption that 2 bags of grain can be produced with the same resources required to produce 1 bag of groundnuts. This rate of transformation is constant, whether 1 bag of groundnuts or 30 bags of groundnuts are produced.

The convex line implies that as more groundnuts are produced, the advantages of increased scale are so great that the sacrifices of grain output per additional bag of groundnuts become smaller and smaller. At no stage are the two enterprises complementary or supplementary to each other.

Where the production possibility frontier is a straight line or convex to the origin, the maximum-profit position will almost certainly be found where all resources are concentrated in one enterprise. The advantage will be in favour of specialization. This may be illustrated by referring to Figure 4.3. For the straight line, the rate of product transformation is constant. Each bag of groundnuts is produced at the cost of 2 bags of grain. Now if groundnuts are priced at £3·50 per bag and grain at £1 per bag, then it would pay to specialize in groundnut production. Each bag of groundnuts produced yields an extra revenue of £3·50 at an opportunity cost of $2 \times £1 = £2$. Thus marginal revenue exceeds marginal cost for every bag of groundnuts produced. It therefore pays to specialize in groundnut production.

In the diagram the iso-revenue line for an inverse price ratio of 3·5 is shown broken. In this case the maximum iso-revenue line just touches the production possibility frontier at one end; namely where all resources are devoted to groundnut production.

If the price ratio changes, the optimum allocation of resources between groundnuts and grain is unaffected until the inverse ratio falls below 2. If groundnuts were less than twice as valuable per bag than grain then it would pay to devote all the available resources to grain production.

Hence, where some enterprises can be expanded to scale and where there are constant or increasing returns to scale it may be more profitable to specialize. However, apart from the possibility of decreasing returns there are also supplementary and complementary relationships between enterprises which may favour diversification.

DIVERSIFICATION

Supplementary enterprises do not compete for certain resources, either because they use different types of resources or because they use the same resources but at different times and in different ways. Even where there are constant or increasing returns to scale in two enterprises they may still be supplementary. For instance maize grown during the early rains and groundnuts grown during the late rains are supplementary enterprises. They hardly compete for any resources since they are grown at different times of the year. Where enterprises are supplementary it will usually be more profitable to produce both rather than one alone. In this way the fixed costs of the resources used will be spread over a larger total output. The introduction of supplementary enterprises is an important method of supplementing farm income, thereby increasing productivity of resources. Hence supplementary relationships should always be borne in mind in farm planning.

Whereas crops grown in different seasons may be supplementary for practically all resources, more often enterprises are supplementary for a particular resource. For instance, different crops grown in a mixture may be supplementary in their use of space or light. In Tanzania, sisal, a prickly crop, is grown in widely spaced rows for ease of access. Cotton grown between the rows does not compete for space or light. In the Sudan in a system of ridge cultivation, maize is planted on the crests to make full use of the light while sweet potatoes are planted on the outward slopes and trail between the ridges. Then again different crops grown in mixtures or in rotations may be supplementary in their use of nutrients. Maize takes up different nutrients from yams. Cereals generally extract nitrates from the soil while root crops make heavy demands on phosphates.

To some extent livestock may be supplementary to crops in the use of land. For instance cattle grazed between the rows of an oil palm plantation are supplementary to oil palms in the use of land. Animals which are fed on crop waste products or on a grass ley which replaces an unproductive fallow are also supplementary in the use of land.

The possibilities for supplementarity in the use of labour are particularly important. The labour requirements of almost all crops have very definite seasonal peaks at the planting and harvesting seasons and for weeding. Hence, if only one crop is grown, the demands for labour will be very variable between different times of the year. If several crops are grown, it is unlikely that the work peaks will coincide, so the total demand for labour is likely to be more even throughout the year. This means that the different crops are supplementary in their use of labour at different times of the year.

Finally enterprises may be supplementary in their use of capital. If a farmer produces only one specialized crop he may be paid for it only once a year, whereas with several enterprises his income will be more regular. With a regular income he requires less capital and is less likely to get into debt than if he received his income all at one time of the year. Similarly failure of a single crop might cause a farmer to get into debt. But it is less likely that several different crops would all fail in the same year.

Complementary relationships are less important in practice but are generally found where the products of one enterprise are used in the production of another enterprise. This is sometimes known as a 'vertical relationship' between enterprises, the discussion so far having been concerned with 'horizontal relationships'. Where there is a vertical relationship between enterprises, the expansion of one may permit the expansion of the other. One example often quoted is the complementary relationship between legumes, such as cowpeas, and other crops. Legumes actually add nitrogen to the soil and this nitrogen may increase the yields of other crops. Within limits, the introduction of a legume into a crop mixture or a crop rotation may lead to an increase in the output of the other crops. Insofar as growing a rotation of different crops gives opportunities for control of weeds, pests and diseases, the different crop enterprises are complementary to one another. The control of pests in one crop is a by-product of the other crop. Within limits expansion of one crop can increase production of the other.

However, no two enterprises are ever complementary over all possible combinations of the two. The complementary relationship always gives way to competition. Similarly most supplementary relationships only apply over a limited range and eventually give way to competition. But if a pair of enterprises are supplementary or complementary over even a limited range, the rate of product transformation of one enterprise for the other must increase eventually. This means that the production possibility curve must bulge outwards; in which case a combination of the two products is likely to be more profitable than specialization on one only. In fact, so long as the enterprises compete for different resources at different rates, the production possibility curve will be concave to the origin and the most profitable system will involve both enterprises (see Chapter 16 on linear programming).

Again this argument will apply where many enterprises are considered. So long as there are supplementary or complementary relationships and increasing rates of transformation, a combination of enterprises will be more profitable than a single specialized enterprise. This will not be the case however if supplementary and complementary relationships are relatively unimportant and there are constant or increasing returns to scale in a particular enterprise.

RESOURCE ALLOCATION AND THE WHOLE FARM

For a single enterprise the least-cost method of production occurs where the marginal product per £1 spent is the same for each resource. For a single resource used on several enterprises, the maximum profit occurs where the marginal value product is the same in each enterprise. These two rules may be brought together to provide a general recommendation for maximizing profits, when a limited sum of money can be spent on various resources to be used on various enterprises: the marginal value product per unit of expenditure should be equal for all resources in all enterprises.

If, however, there are no constraints on the resources available, then the use of each resource should be expanded up to the economic optimum in each enterprise; marginal cost should be equated with marginal revenue. Just as the individual enterprise may be expanded to scale so too may the whole farm. In fact because of supplementary and complementary relationships, the advantages of expanding a single enterprise to scale may be limited. This qualification does not apply if the whole farm, or combination of enterprises, is expanded to scale. As we have seen, if there are constant or increasing returns to scale there is no economic optimum. However, even if the whole farm is expanded it is likely that fixity of some resource, such as management or land, will eventually lead to decreasing returns.

In fact we can summarize all this into a single general rule for rational economic behaviour. For this purpose we must consider all technical relationships to be technical substitutions of one good or service for another, whether a factor is substituted for a product, one factor is substituted for another factor or one product is substituted for another product. Similarly we must consider all inverse price ratios to represent the rate of value substitution based on individual choice. Then the economic optimum or equilibrium point occurs where the rate of technical substitution is equal to the rate of value substitution. If a man values his leisure at 1·8 lb. of groundnuts per hour but technically he can transform leisure into 2 lb. of groundnuts per hour he should do so until the rate of technical substitution falls to 1·8 lb. The same principle can be used when substituting leisure for savings or groundnuts for grain.

SUGGESTIONS FOR FURTHER READING

DILLON, J. L. (1968) *The analysis of response in crop and livestock production*, Oxford, Pergamon Press.
DOLL, J. P., RHODES, V. J. and WEST, J. G. (1968) *Economics of agricultural production, markets and policy*, Homewood, Illinois, Irwin.
HEADY, E. O. (1952) *Economics of agricultural production and resource use*, Englewood Cliffs, Prentice-Hall.
VINCENT, W. H., ed. (1962) *Economics and management in agriculture*, Englewood Cliffs, Prentice-Hall.

Mathematical appendix to Chapter 4

THE PRODUCTION FUNCTIONS

Let us assume that only one variable resource must be allocated between two products (enterprises). The quantities of other resources used in the production of each are fixed. Thus we have two functions: $Y_1 = f_1(X_1/X_2, X_3, \ldots, X_n)$ for product (1) and $Y_2 = f_2(X_1/X_2, X_3, \ldots, X_n)$ for product (2).

The variable resource X_1 may now be called X because there is no need to distinguish it from other inputs. It has two average products,

$$APP(Y_1) = \frac{Y_1}{X}$$

and

$$APP(Y_2) = \frac{Y_2}{X}.$$

Similarly there are two marginal products,

$$MPP(Y_1) = \frac{\Delta Y_1}{\Delta X}$$

and

$$MPP(Y_2) = \frac{\Delta Y_2}{\Delta X}.$$

The rate of transformation to product (2) from product (1) is defined as

$$RPT_{21} = -\frac{\Delta Y_1}{\Delta Y_2}.$$

For competing enterprises an increase in Y_2 must result from a decrease in Y_1 so

$$\frac{\Delta Y_1}{\Delta Y_2}$$

is negative and RPT_{21} is positive.

If inputs of the variable resource in enterprise (1) are increased by ΔX then the output of product (1) should increase by ΔX times its marginal product:

$$\Delta Y_1 = \Delta X \,.\, MPP(Y_1).$$

Dividing both sides of this equation by $MPP(Y_1)$ gives

$$\Delta X = \frac{\Delta Y_1}{MPP(Y_1)}.$$

For total inputs X to remain constant, an increase of ΔX in enterprise (1) must be associated with a decrease of ΔX in enterprise (2), so

$$\Delta Y_2 = -\Delta X \,.\, MPP(Y_2)$$

and by dividing both sides by $-MPP(Y_2)$

$$\Delta X = \frac{-\Delta Y_2}{MPP(Y_2)}.$$

Since ΔX is the same in both cases

$$\frac{\Delta Y_1}{MPP(Y_1)} = \frac{-\Delta Y_2}{MPP(Y_2)}.$$

Multiplying both sides by $-\dfrac{MPP(Y_1)}{\Delta Y_2}$

gives

$$-\frac{\Delta Y_1}{\Delta Y_2} = \frac{MPP(Y_1)}{MPP(Y_2)}.$$

But

$$-\frac{\Delta Y_1}{\Delta Y_2} = RPT_{21} \quad \text{so} \quad RPT_{21} = \frac{MPP(Y_1)}{MPP(Y_2)}.$$

ENTERPRISE COMBINATION

Now let

P_X = unit factor cost of X_1,

$P(Y_1)$ = price per unit of Y_1,

$P(Y_2)$ = price per unit of Y_2,

$MVP(Y_1)$ = marginal value product of X in producing Y_1,

and $MVP(Y_2)$ = marginal value product of X in producing Y_2.

The profit-maximizing combination of enterprises may now be expressed in several ways (note similarities with least-cost combination of resources).

1. Profit is maximized where the extra return from expanding production of enterprise (1) just equals the extra cost in terms of reduced product of enterprise (2) (opportunity cost):

$$\Delta Y_1 . P(Y_1) = -\Delta Y_2 . P(Y_2). \tag{1}$$

2. Dividing both sides by $-\Delta Y_2 . P(Y_1)$ gives

$$-\frac{\Delta Y_1}{\Delta Y_2} = \frac{P(Y_2)}{P(Y_1)}$$

but

$$-\frac{\Delta Y_1}{\Delta Y_2} = RPT_{21}$$

therefore

$$RPT_{21} = \frac{P(Y_2)}{P(Y_1)} \tag{2}$$

i.e. rate of product transformation equals the inverse price ratio.

3. But

$$RPT_{21} = \frac{MPP(Y_1)}{MPP(Y_2)}$$

hence by substituting in equation (2) this gives

$$\frac{MPP(Y_1)}{MPP(Y_2)} = \frac{P(Y_2)}{P(Y_1)}.$$

Multiplying both sides by $P(Y_1) . MPP(Y_2)$ gives

$$MPP(Y_1) . P(Y_1) = MPP(Y_2) . P(Y_2)$$

or

$$MVP(Y_1) = MVP(Y_2) \tag{3}$$

i.e. marginal value product per unit of a variable resource should be equal in each enterprise: *principle of equal marginal returns.*

4. Equation (3) is consistent with the rule for determining the economic optimum level of resource use, which states

$$MVP_X = P_X$$

(see Mathematical appendix to Chapter 2). Hence

$$MVP(Y_1) = P_X = MVP(Y_2)$$

(compare with equation (3)).

5 Price variation, risk and uncertainty

OUTPUT AND PRICES

The assumption that a farmer can sell any amount of his produce at a constant fixed price is only justified if the quantity he sells represents a small proportion of the total. A single cocoa grower producing between 1 and 2 tons out of a total national product of 200,000 tons of cocoa is unlikely to influence the price per ton by varying the amount he sells. However, a farmer growing maize for sale in a small village market may produce a significant proportion of the total quantity of maize sold there, which means that he is in a position to influence the local price of maize by varying his output. This would be a case of 'imperfect competition'. In an extreme case there may be one single producer of a particular commodity, such as eggs, in the whole community who would therefore have a pure 'monopoly' in the sale of his product. A large unit such as a producers' marketing cooperative, a farm settlement, a state farm or a plantation is more likely to be in a monopoly position than an individual smallholder.

A producer selling under conditions of imperfect competition can only increase his sales if consumers will buy more of his product, but they are only likely to do so if he reduces the price. This is because, generally speaking, the more a person has of one commodity, the less he values it in relation to other commodities and hence the lower the price he is prepared to pay. Furthermore different people have different incomes and different tastes. As the price of a commodity falls, more of the poorer people will be able to buy some of it, in addition to the richer people buying more of it. Thus if the price of eggs falls not only may adults eat more eggs but also children may be given eggs for the first time.

Where increasing output results in falling product price the economic optimum still occurs where marginal cost equals marginal revenue *but* marginal revenue is now less than product price. This is because the sale of an extra unit of product causes a fall in the price for every other unit of output, as shown in Table

5.1. As output is increased and the price falls, so too does the marginal revenue, so that even where there are constant returns to scale and hence a constant marginal cost per unit of output an economic optimum is reached when the marginal revenue falls to the level of the marginal cost.

This is broadly similar to the situation on a subsistence farm where the entire output is used for home consumption. As the output of a single commodity, say maize, is increased, the family

TABLE 5.1 Marginal revenue under imperfect competition

| | pence | | |
Dozens of eggs offered for sale	Price per dozen	Total revenue	Marginal revenue per dozen
1	30	30	30
2	28	56	26
3	25	75	19
4	21	84	9
5	16	80	−4

will tend to value each bag of maize lower in relation to other commodities and in relation to their own labour inputs. The economic optimum occurs where the subjective 'marginal cost' equals the subjective 'marginal revenue'.

Likewise there may be imperfect competition in the hire or purchase of resource inputs. Indeed some inputs cannot be hired or bought and are therefore fixed. Others may be controlled by one or a few individuals who can therefore influence the price by limiting the amount available. For instance a single fertilizer manufacturer or importer would probably have to accept lower prices the more he tried to sell. By restricting his sales, however, he could force the price of fertilizer up above the marginal cost per ton.

Imperfect competition may result where there is only one or a few *buyers* of a product or an input. A single buyer is said to operate under conditions of 'monopsony'. A farmer who is the sole employer of labour in the district can only increase his inputs of hired labour if he increases the wage he offers per unit of work. This is because, generally speaking, the less leisure each worker has, the more he values it in relation to other forms of consumption and hence the higher the wage he asks. Furthermore the higher the wage rate, the more people will be persuaded to work.

Where increased inputs of a given resource are associated with increasing price per unit of the resource, the economic optimum still occurs where the extra cost of using one more unit is equal to the marginal value product, *but* now the extra cost is higher than the unit factor cost. This is because the purchase of an extra unit of input causes a rise in the price for every other unit of input. Again a similar situation may arise where production is dependent upon family resources. As the family work more hours the subjective cost per hour of work is likely to increase. As the family divert more resources to creating capital assets, the subjective cost per unit of capital is likely to increase.

Thus there may be many situations in which the prices farmers receive for their products and the prices they pay for their inputs vary with the level of output. The general rule defining the economic optimum, that is the point where the rate of technical substitution is equal to the rate of value substitution, still applies but the price variation must be taken into account. In Chapters 2, 3 and 4 we simplified the analysis by assuming that prices and hence rates of value substitution are constant. In practice both the rate of technical substitution and the rate of value substitution may change as one commodity is substituted for another. This is illustrated by the graphs in Figure 5.1, which give the economic optimum when it is assumed that the rate of value substitution varies. They may be compared with Figures 2.4, 2.9 and 4.2 which illustrate the economic optimum when there is a constant price ratio.

QUALITY AND PRICE

The price obtained for many farm products depends upon the quality. Women will pay more for rice which is free of dirt, stones and insects, and this is true for most food crops; large eggs sell for a higher price than small eggs; fat animals sell for a higher price than bony animals. Where there is a marketing board for export crops there is usually an official system of grading on the basis of quality standards, with a scale of prices depending upon the quality grade. Crops commonly graded for sale include tea, coffee, cocoa, tobacco, cotton, rubber, groundnuts and palm oil.

Some of the factors affecting quality and grading are evenness of sample size or shape, condition (which may depend upon maturity or decay), freedom from foreign matter, flavour in the case of food products and freedom from disease and insect damage.

(*a*) **Factor-product**

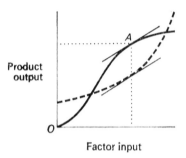

Product
output

Factor input

(*b*) **Factor-factor**

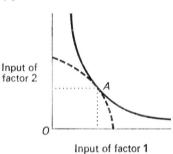

Input of
factor 2

Input of factor 1

(*c*) **Product-product**

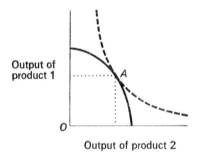

Output of
product 1

Output of product 2

FIGURE 5.1 The economic optimum under imperfect competition. The point of maximum profit is defined in the same way as under pure competition but the lines joining points of equal value are now curved as rates of value substitution vary.

Product quality can be improved to some extent simply by more careful management but major improvements may involve the use of new varieties, more labour per unit of output or new equipment, any of which would increase costs. It is only worth making the effort to improve the quality of farm produce if the gain in price per unit exceeds the extra cost. Again there is an economic optimum level of quality at which to aim, where the marginal cost of further improvement is just equal to the marginal benefit.

PRICE CHANGES OVER TIME

Although individuals, by their own actions, may have no perceptible influence on market prices under pure competition, the combined decisions of a large number of individuals may bring about sizeable price changes. For example an individual farmer may plant a new variety of maize and increase his production without affecting the market price. But as large numbers of other farmers adopt this new variety, total maize production will be increased so that prices will be affected. This is but one example of the way new production techniques can lower costs, increase total output and thus lead to a fall in product price. In a similar manner the decision of one family to buy a ram to celebrate Id al-Fitr would have little effect upon the market price. But if many families buy rams for the same occasion there will be a marked temporary rise in price.

There are many other forces which contribute to price fluctuations for farm products. They operate by causing changes in either the quantity supplied for sale or the quantity desired for purchase or both. A general increase in the quantity offered for sale is likely to result in a fall in price, other things being equal; conversely a general increase in the quantity desired for a purchase is likely to result in a price rise.

Four types of price movement may be distinguished. First there are long-term trends normally lasting over a period of several years or longer. These are caused by changes in population, incomes, tastes and techniques of production. Second, some prices move in cycles: alternating rises and falls every few years. To some extent these cycles may be self-perpetuating. The third type is a seasonal fluctuation in prices which takes place within a year and may result from the seasonal pattern of production, prices tending to fall at harvest time. Alternatively a seasonal price change may be caused by variation in the quantity desired by consumers. Thus the price of poultry and other meat rises at the time of major festivals during the year. Finally there are irregular movements many of which may be quite unpredictable. They may be caused by a drought, a sudden outbreak of pests or diseases of crops or livestock or they may result from an outbreak of war, an influx of refugees or some natural catastrophe.

From the farmer's point of view there is an important distinction between those price changes which can be predicted and those which cannot. He can allow for predictable price changes in making his management decisions but unpredictable changes are a cause of risk and uncertainty. Most seasonal price fluctuations are predictable so a farmer may plan the timing of his production to coincide with periods of high prices. By harvesting and selling his crops early, before most other farmers, he might avoid the low prices at harvest time. This means that he must plant early. Alternatively he may store his crops at harvest time to sell or consume them later when market prices have risen. The production of poultry and other livestock may be timed so that they are ready for slaughter and sale just before a major festival such as Christmas or Easter, Id al-Kabir or Id al-Fitr, or the national day of independence.[1] Naturally, if large numbers of farmers change their pattern of production so as to take advantage of seasonal price increases, this in itself will tend to reduce these price increases, so the benefits of adjusting to these seasonal increases will be smaller. However, it is most unlikely that the seasonal price differences will be eliminated altogether because costs are involved in early planting, storage or careful timing of production. There would be no incentive for farmers to adjust if the differences in product price were insufficient to cover the extra costs.

Price trends and cycles are less easy to predict although careful study of price movements in the past can provide a basis for estimating how they will move in the future. However, irregular price changes are by their very nature largely unpredictable. Farmers who sell their products and who buy or hire inputs must take their managerial decisions under conditions of uncertainty about future prices. They face the risk that the decisions they take will turn out to be inappropriate to the new price situation which has arisen and was not foreseen when the decisions were taken. In other words they face the risk of losses because of unforeseen price movements. However, unpredictable price changes are not the only causes of risk and uncertainty. Indeed there is considerable risk and uncertainty in pure subsistence farming.

RISK, UNCERTAINTY AND SUBSISTENCE FARMING

A distinction is sometimes made between risk and uncertainty, the former referring to situations in which the outcome is not certain, but where the probabilities of the alternative outcomes are known, or can be estimated. For instance, if a farmer knows that his millet crop is likely to fail one year in four, then this is a case of risk. He does not know whether the crop *will* fail in any particular year but he knows that over a long period it is likely to fail one year in four on average. Alternatively if the farmer has no idea of the probability of crop failure he is in a state of uncertainty. However, the distinction is not very clear in practice and the words are often used interchangeably to mean roughly the same thing.

All production implies the taking of some risk, since the cost or

[1] See Olayide, S. O. and Ogunfowora, O. (1970) Optimal timing in broiler production: an economic analysis, *Bulletin of Rural Economics and Sociology*, Ibadan, vol. 5, no. 2, p. 277.

the effort has to be incurred in advance of the final output, which may turn out to be more or less than was expected. Crop and livestock yields are highly dependent on the weather; rains may arrive early or late or there may be too much or too little rain for normal plant growth. In periods of drought the restricted plant growth means that livestock fodder supplies and hence livestock production may also be limited. Then again outbreaks of animal or insect pests or diseases can cause major yield losses. Apart from the risks of crop failure or losses of livestock production there are also risks of sickness, injury or death of family members. Collinson reports the effect of the farmer falling ill on a trial unit farm operated by a local family in Tanzania. 'On 19th December he finalized his cotton planting for the season with a total of 3·21 acres, still with adequate time to complete his remaining cultivations on schedule. On Friday 21st he went down with malaria for a full week, the whole family stopped work to care for him.'[2] This incident reduced the labour supply in the critical month for cultivation and planting by 27 per cent and the total farm output by a similar amount.

Many subsistence farmers produce such small surpluses over the basic nutritional needs of their families that any crop failure can jeopardize their very survival. For this reason they may be particularly averse to taking risks and therefore prefer more reliable enterprises and methods.

Indeed some would argue that the main objective of subsistence producers is survival rather than the maximization of the surplus of benefits over costs. However, the two are not necessarily in conflict; the greater the output a family produces with its limited resources or the lower the cost a family incurs in producing a given output, the more is that family able to survive poor cropping seasons or other catastrophes. Food surpluses can be stored to tide the family over periods of shortage and unused labour may be called upon in times of stress. Nevertheless farmers generally take some precautions against risk, which may limit their total output or increase their total costs over a period of years.

The customs and organization of traditional society provide the individual family with a measure of security against risk. There is generally an obligation on the more fortunate and able members to help their kinsmen or neighbours in times of need. This may relieve the situation in cases of sickness, injury or death but a failure

[2] Collinson, M. P. (1969) Experience with a trial management farm in Tanzania, *East African Journal of Rural Development*, vol. 2, no. 2, p. 28.

of the rains or an attack of crop pests may affect the whole community so that all are in the same boat. Each individual farm family needs to take some precautions against risk. These include (1) choice of reliable enterprises, (2) diversification, (3) keeping reserve food supplies, and (4) maintaining flexibility.

All localities have some crop or livestock enterprises which are more reliable than others. The returns from some enterprises vary greatly from year to year; they give large yields in some years and in other years fail completely. Other enterprises are more stable; while yields may vary from year to year, the ups and downs are not extreme. The more reliable enterprises or varieties of crops and livestock may be chosen in preference to the one with variable yields even when the latter gives a higher return on average. Thus farmers may choose a drought-resistant variety of sorghum or millet in preference to a variety which is higher yielding on average but liable to fail in drought conditions. When a new variety is introduced, although the yields may be less variable than those of traditional varieties, the farmers have no experience of growing it so there is greater risk involved in trying the new variety. This may explain why farmers are reluctant to use new and apparently improved varieties.

Diversification is a common precaution taken to meet risk. Several different crops are grown or livestock kept in the hope that they will not all fail together. If one should fail the income from the other enterprises may be sufficient to keep going. The farmer avoids putting 'all his eggs into one basket'. Clearly diversification only has the effect of reducing risk if the yields of the various enterprises are not positively related. If they are positively related and when one fails they all fail, diversification does not reduce risk. Intercropping is a common form of diversification in traditional farming practice. Crops which are more resistant to drought are planted together with preferred foods to ensure that some return is obtained from the effort put into land preparation. Sorghum and maize are grown together in parts of Africa, sorghum being drought-resistant but susceptible to bird damage while maize is liable to fail in a drought but is more resistant to bird damage. A farmer may also diversify by planting the same crop at different times or on different soil types. Staggered planting may be necessary to ensure an even supply of food over as long a period as possible, but it also reduces risk since periods of water stress will occur at different stages in the growth of crops planted at different times. While some plantings may suffer, those with relatively low water needs at that time are more likely to survive. By planting crops on different soil types the farmer diversifies to avoid risk: in a dry year the crop on sandy, upland soils may fail, in a wet year the crop on wet, river-valley land may fail. The opportunity offered for reducing risk by growing crops on various types of soil may be one reason why farmers choose to operate fragmented farms.

Food reserves also provide some security against the risk of crop failure, but stored foods are liable to deterioration and losses. Cassava has the particular advantage that it can be left in the ground for up to two years without deterioration; indeed it continues growing. This makes cassava a particularly useful famine reserve in many parts of Africa. Finally the farmer will avoid risk by maintaining flexibility. This means he is prepared to make new decisions from day to day in the light of changing conditions. He does not make a firm plan at the start of the growing season and stick to it rigidly. Thus if an early-planted staple food crop fails he may replant with a more drought-resistant variety. If he feels none will be successful he may increase the area of famine-reserve crop.

RISK PRECAUTIONS IN COMMERCIAL FARMING

In addition to the risks, already mentioned, of yield variation or crop failure and sickness, injury or death of family members, commercial farmers who sell products and hire resources face the risks associated with unpredictable price variations and uncertainties about the behaviour of others with whom they have commercial dealings. Commercial farmers also take precautions against risk and uncertainty so subsistence farmers are not unique in this respect. Indeed a desire for security is a normal motive of all human beings.

This means that farmers like other people are risk avoiders; they are prepared to forgo a certain amount of income every year in order to avoid occasional large losses.[3] One method of risk avoidance is to discount for risk; that is actually to produce less than the economic optimum level of output every year in order to reduce losses in bad seasons. We may illustrate this with the hypothetical example from Chapter 2. The figures are taken from

[3] People who enjoy gambling are exceptions, but most of them would not be prepared to gamble with a very large proportion of their total income over a given period.

Table 2.1 and 2.2. unaltered but now they are assumed to represent the relationship between average expected inputs and outputs, rather than an exact unchangeable relationship between actual inputs and outputs.

Now we assume that the value of the total product can vary by 50 per cent either side of the average as shown in Table 5.2. This means that in bad years the farmer makes losses but these are counterbalanced by larger than average profits in good years. Over a period of years profit is still maximized by using about 4 units of labour as shown in Chapter 2. However in bad years, the use of this amount of labour would result in a loss of £11·50. The loss in the bad years is minimized at £9·50 by using only 3 units of labour. Some farmers would prefer to use only 3 units of labour every year in order to minimize losses in bad years even though this would mean forgoing a certain amount of profit (in this case £7 — £6 = £1) on average. If it were possible to choose an input of between 3 and 4 units of labour the majority of farmers would probably choose a level of input somewhere between that which would maximize average profit and that which would minimize losses in bad years.

The choice between profit maximization and security is particularly obvious where it is possible to insure against risks. By such insurance a private company or state organization guarantees to pay a substantial sum in the event of a major catastrophe in return for a relatively small annual premium. Insurance may cover major risks such as the death of the farmer or some member of his family; it may be used for sickness and accidents which disable the farmer, and for fires or other hazards which can destroy capital items such as buildings, breeding and fattening stock, cars, lorries and other machines. In some countries it is possible to insure against the loss of crops by hail or hurricanes. Clearly farmers and others who buy insurance forgo a certain amount of income each year (the annual premium) in return for the security offered. Insurance is not common among African farmers except perhaps where tractors and other large machines are used.

Although no one can predict the future exactly, the more knowledge a farmer has about the relevant production relationships and price movements in the past, the more accurate he is likely to be in predicting the future. Thus a farmer can reduce risk and uncertainty by collecting information about the costs and returns for the alternatives open to him and about market prices. A farmer who keeps a record of market prices is more likely to be able to

TABLE 5.2 Risk and the economic optimum

£

I	II	III Bad years			IV Good years			V Average		
Units of labour input	Total cost	Value of total product	Profit	Value of marginal product	Value of total product	Profit	Value of marginal product	Value of total product	Profit	Value of marginal product
0	10	0	−10		0	−10		0	−10	
				3·50			10·50			7
1	15	3·50	−11·50		10·50	−4·50		7	−8	
				6·50			19·50			13
2	20	10	−10		30	10		20	0	
				5·50 —3			16·50 —9			11
3	25	15·50	−9·50		46·50	21·50		31	6	
				3			9			6
4	30	18·50	−11·50		55·50	25·50		37	7 —5	
				1·50			4·50			3
5	35	20	−15		60	25		40		

Note. Average results are taken from Table 2.2. Now total product is assumed to be 50 per cent lower than average in bad years and 50 per cent higher in good years. Cost per unit of variable input is assumed to remain constant at £5. Note that a negative profit is a loss, so profit is maximized or loss minimized at the economic optimum, where the value of the marginal product is approximately equal to the unit factor cost.

predict future price trends and cycles than his less well informed neighbours.

Price uncertainty could be eliminated entirely if the farmer could make advance contracts with the buyer of his products and the sellers of his inputs. Unfortunately, the opportunities to make contracts for what the farmer buys and sells are limited, although some opportunities do exist. For instance a livestock-feed mill may contract to buy farmers' grain at a price agreed in advance, a tobacco company may do the same for the tobacco crop. For many cash crops in different parts of Africa, the marketing boards provide the same service by announcing the prices at which they will buy produce, in advance. Payments for land and hired labour are generally fixed by contract.

Commercial farmers also take the same risk precautions as subsistence farmers do. They prefer the more reliable enterprises, with yields and prices that do not vary much from year to year, though these may not be the most profitable. One reason why farmers continue to grow food crops in areas suited to very profitable cash crops may be that they consider the returns from food crops to be more reliable. This is also an instance of diversification to avoid risk. Another example of diversification is found where farmers both cultivate arable crops and keep cattle as in the semi-arid areas of Botswana and Sukumuland. In periods of drought when the crops fail, cattle are sold. This may be necessary because fodder is in short supply but it does mean that the farmers have some cash income with which to buy food for themselves and their families. Thus by diversifying and keeping both crops and livestock farmers reduce the risk of going hungry.

In an exchange economy, farmers also carry reserves as a precaution against risk. Indeed the cattle in the areas just mentioned represent one kind of reserve. Cash savings are the most flexible kind of reserves, since they can be used to meet almost any kind of need. A farmer who depends upon cattle grazing for his income may wish to keep some cash in reserve to buy fodder for his cattle if a drought should occur. Reserve capacity is also provided by employing more labour or machinery than is needed in an average year, in order to be prepared for the occasional years when cultivation conditions are difficult.

Farmers who sell cash crops may reduce risk by maintaining a flexible system of farming so that they can change to another enterprise if the price of the main cash crop falls. Specialist tree crop production is a highly inflexible system since it certainly cannot be

changed easily and quickly. From this point of view oil palms may
be less risky than many other tree crops, because oil palms provide several alternative products. If the price of palm oil falls,
the trees can be felled for palm wine tapping or for building
materials. The construction of a grain store increases flexibility
in one respect and decreases it in another. It increases flexibility in
that the farmer can sell his grain when it suits him or when prices
are high, whereas without the store he is forced to sell at harvest
time. It decreases flexibility because a grain store is useless for any
productive use other than storing grain. Generally a farm system
is kept flexible by using temporary hired labour rather than regular
workers, by using any kind of labour rather than specialized
machinery or by hiring machinery rather than buying it.

In summary, farmers may use some or all of these precautions
against risk:

(1) choice of reliable enterprises;
(2) diversification;
(3) keeping reserves;
(4) maintaining flexibility;
(5) discounting for risk;
(6) insurance;
(7) collecting more information;
(8) contracting prices in advance.

All of these are likely to incur a certain cost or reduction in
profit, but this must be set against the benefit of greater security.

SUGGESTIONS FOR FURTHER READING

ANDREWS, P. W. S. (1964) *On competition in economic theory*, New York, St. Martin's.
GOULD, P. R. (1963) Man against his environment: a game-theoretic framework,
Annals of the Association of American Geographers, vol. 53, no. 3, p. 290.
GOULD, P. R. (1965) Wheat on Kilimanjaro: the perception of choice within game
and learning model frameworks, *General Systems Year Book*, vol. 10, p. 157.
LINDLEY, D. V. (1971) *Making decisions*, London, Wiley: Interscience.
NICHOLLS, W. H. (1941) *A theoretical analysis of imperfect competition with special
application to the agricultural industries*, Iowa State University Press.
OFFICER, R. R. and ANDERSON, J. R. (1968) Risk, uncertainty and farm manage-
ment, *Review of Marketing and Agricultural Economics*, Sydney, vol. 36, no. 1, p. 3.
ROBINSON, J. (1933) *The economics of imperfect competition*, London, Macmillan.

Part 2 RESOURCE USE

6 The environment and land use

THE ENVIRONMENT

Although it is probably true to say that every farm is unique in certain respects, there are many similarities between large numbers of farms over quite large areas. In any one such 'type of farming area' or 'land-use zone', the same basic foods and often the same cash crops are grown by practically all the farmers; methods of production are broadly similar and so too are attitudes, customs and social institutions. Thus, although there may be varia-tions in farm sizes, and in some of the minor crops grown and live-stock kept, the same basic system of farming is found over the whole area.

The system of farming found in any particular area is determined by the following features of the environment:

(1) density of agricultural population;
(2) natural resources—the physical environment;
(3) location in relation to markets, roads and railways;
(4) institutions relating to the land—land tenure;
(5) technical knowledge and capital resources available.

This classification is somewhat arbitrary, since, as we have seen, it is not always possible to distinguish between natural resources and the man-made improvements which increase the total pro-duct per hectare; a particular system of farming may itself either improve the structure of the soil, or damage it through erosion and exhaustion, thus altering the physical environment.

Nevertheless, the management decisions of African farmers are influenced by these five main features of the environment, which, as individuals, they can do little to change. For instance an indi-vidual farmer may be free to choose the location of his home and farm land, but only within certain quite strict limits set by the pattern of land tenure. He may improve his technical knowledge and introduce new forms of capital onto his farm but again he is limited to those resources that are available in his area. It is very difficult for a man to start growing a new crop, say soya bean, if there is no seed and no knowledge of how to grow it available in the area. It is very difficult for a man to start using tractors in an area where there are no tractor service stations and no one with knowledge of tractor driving and maintenance. A large farmer or the government introducing new knowledge and capital into an area is providing 'external economies' for the small farmers in the area. A road, a land reclamation scheme or an agricultural research station provided by the government yields external economies to individual farmers, unrelated to the amount of taxes they pay. Farmers may also suffer from dis-economies, when their land is taken for the construction of a new dam, or a main road is driven through the middle of their farms.

The first four features of the environment mentioned above and their influence on farm systems will now be discussed in more detail. The effects of technical knowledge and capital resources are so bound up with the other features that they will not be discussed separately, though we shall return to the subject of capital in a later chapter.

POPULATION DENSITY

The density of population engaged in farming has an important influence on the system adopted in any area, since it controls the average amount of land available to the farm families and hence the intensity of farming. A Danish economist, Mrs. Boserup, has indeed argued that the growth of population is the main force by which agricultural development is brought about since intensive farming often involves more advanced techniques and more capital per hectare and per family than extensive farming.[1] She classifies land-use systems into five main types, listed here in order of increasing frequency of cropping and hence increasing intensity of labour inputs per hectare of all land under both cultivation and fallow.

(1) Forest-fallow cultivation, where land is left for as many as twenty-five years to regenerate forest, after a year or two of cropping.

[1] Boserup, E. (1965) *The conditions of agricultural growth*, London, George Allen and Unwin.

(2) Bush-fallow cultivation, where the fallow period extends from about six to ten years, to allow the regrowth of small trees and bushes.

(3) Short-fallow cultivation, based on natural grasses, with cultivation recurring after intervals of perhaps one or two years.

(4) Annual cropping of land, which may be left fallow for some months or weeks only, between the harvesting and planting of annual crops.

(5) Multi-cropping of land, with each plot bearing two or more crops each year and virtually no fallows.

These land-use systems are claimed to be associated with particular techniques of production. The system of forest fallow is associated with primitive techniques using neither hoe nor plough. Secondary forest is cleared by burning and seeds are sown in the ashes using a simple digging stick. When forest fallow is replaced by the shorter bush fallow, the hoe is necessary to clear the bush and grasses which are not destroyed by burning. The plough is needed for short-fallow or continuous cultivation because grass roots are exceedingly difficult to remove by means of hoeing. Furthermore the gradual disappearance of roots of trees and bushes in the fallow facilitates the use of the plough. Grass fallows are more suitable for grazing than bush or forest fallow so grazing livestock and mixed farming can be introduced at this stage. At the same time it is argued that when fallow is shortened or even eliminated techniques of fertilization must be introduced in order to maintain crop yields. Likewise the use of irrigation may be necessary for multi-cropping or even annual cropping. Irrigation facilities and other land improvements, such as terracing, are never used with long fallow and rarely with short fallow.

Mrs. Boserup's theory is that farmers know about these more advanced and intensive techniques of production but do not adopt them until forced to do so by population pressure because they involve more and harder work. Forest fallow requires very little effort by the producer, who burns the bush, plants seeds in the ashes and later collects the harvest. Contrast this with the multi-cropping farmer who is busy throughout the year, cultivating, planting, weeding, fertilizing, perhaps irrigating and finally harvesting two or more crops each year. There is no doubt a large element of truth in this idea, which means that as the density of agricultural population and hence the intensity of labour use per hectare increases, the marginal product per unit of labour declines. This general trend is borne out by Miss Haswell's study already described and is an example of the working of the law of diminishing returns. At the same time, increasing intensity of land use is likely to be associated with increasing land value and hence increasing concern over its allocation and conservation.

From an economic point of view, land and other natural resources have two peculiar features: their total amount is fixed, and they cost nothing to produce. These features distinguish them from capital resources which are the result of human effort, and which can be increased at a cost. In a sparsely populated area where land is freely available to all resident families, its opportunity cost is zero; the use of another hectare of land for cultivation does not diminish the output of some other productive enterprise. If labour is limited to the members of farm families we may think of labour and management as fixed factors and land as a variable factor. For each family, the economic optimum level of land use occurs where the marginal product per hectare is zero, since at this point the total product of the family's fixed resources is at a maximum (Figure 6.1). Beyond this point the total product would be reduced by spreading the fixed factors over more land since the marginal product per hectare of land is negative. Although some land may be unused and local farmers may appear wasteful in consequence, it would clearly be nonsense to extend the area cultivated, if the result would be a fall in total product. However as the farming population grows the area of land in use is extended to keep pace, so the marginal products per unit of land and labour stay constant. Extensive systems such as shifting cultivation with long fallows are appropriate.

Once the critical population density is reached, when all the available land in the area is in use under the existing system, there will be competition for extra land; one farmer can only extend the area he farms if some other user gives up some land, and growing more cash crops involves the loss of land used for some other product. Hence each hectare of land now has an opportunity cost, measured by the value of its marginal product in the next most profitable use. Now as the agricultural population increases within a fixed area of land, as the ratio of labour to land increases, marginal and average returns to labour fall, while the marginal and average returns to land increase as shown in Figure 6.2. As the labour force employed increases, the amount of land available per person declines, that is there is a movement backwards from point A towards the origin O; if the rise in population reduces the amount of land per person from OA to OB, the total product per

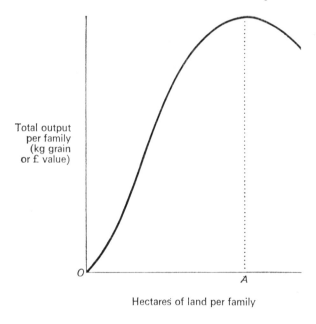

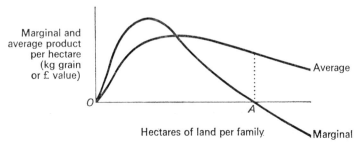

FIGURE 6.1 Response curve for varying land use. If land is freely available, point *A* is the economic optimum as well as the technical optimum.

person falls but the marginal product per hectare of land rises from zero to *OR*. Land now has a value or opportunity cost. It would be worth while for a farmer to hire land for anything up to *OR* kg of grain if this were possible. Even where land is not hired, a farmer considering using land to grow cotton must take this opportunity cost, of *OR* kg of grain, into account.

Where land is freely available and has no cost to its users, cultivators do not normally consider the effect of their own actions upon its future supply. With shifting cultivation, farmers do not spend money or effort on manures, on conserving fertility or on making terraces to restrain erosion; as plots become exhausted, they move to new ones. Similarly, cattle graziers make no effort to improve or to conserve the natural grazing, and in fact over-grazing has caused erosion and devastation of much African grass-land. However, so long as land is available without cost, the diversion of resources to its conservation might well be regarded as wasteful. Such an idea may be strongly opposed by conservation-ists, who often claim that the maintenance of soil fertility is an unchallengeable principle, but it is possible that future generations might benefit more from rapid growth of production even at the

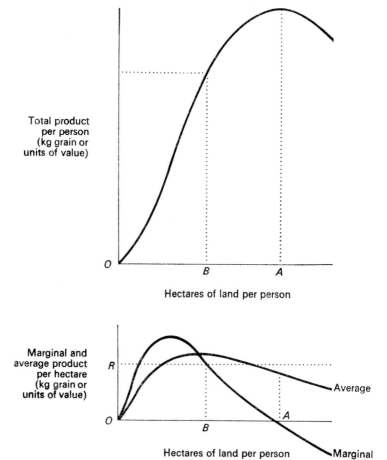

FIGURE 6.2 The effect of increasing population density. As population density increases, hectares per person decrease. As a result total product per person falls and marginal and average product per hectare of land rises.

expense of some natural resources, rather than from investing, in their conservation, factors which have a higher potential yield in other uses.

Some measures of conservation and manuring of soil may however pay individual farmers if clearing is necessary to bring land under cultivation, so that there is a positive cost incurred for each additional unit of land brought under cultivation. Then profits will be maximized by extending the use of land to the point, not where the marginal product per hectare of land is zero, but where it is equal to this cost.

Once all the available land is in use and therefore has a positive value then further effort and expenditure on soil and water conservation and improvement is justified. Thus, in the closely settled zone round Kano, in Northern Nigeria, the land is made to support intensive cultivation partly by irrigation, partly by the use of town wastes as manures, carried out of the city on donkey-back.

The Wakara tribe, on the island of Ukara in Lake Victoria, have evolved a system of intensive cropping based on heavy applications of cattle manure, and the digging in of a leguminous crop which adds nitrogen to the soil. The cattle are kept indoors, fed by hand on elephant grass which is grown as a forage crop, and bedded down on crop residues which retain the manure for use on the plots of cultivated land. Food crops such as bulrush millet, groundnuts and sorghum are grown in rotation with the green manuring crop. This highly intensive system involves a great deal of labour for feeding the cattle, taking the manure to the scattered plots, and for growing and digging in the leguminous crop.

Again, the Wachagga of the southern and eastern slopes of Mt. Kilimanjaro have developed an intensive system of continuous cultivation, based on bananas, coffee, grain crops and cattle, which are again stalled and hand-fed, in order to conserve the manure for use on the cropped land.

Other examples of intensive farming with irrigation or terracing can be found among the Cabrai of Northern Togo, the hill tribes of the Jos area in Nigeria, the Matengo of south-west Tanzania, the hill pagans in Adamawa province of Nigeria, and the Dogon of Mali.

However, the influence of population density in determining the intensity of land use has perhaps been overstated. Cases can be cited where neither the physical environment nor the technical knowledge encouraged greater intensity of land use; increasing population resulted in migration of whole communities, or in the migration of many young men in search of wage employment, leaving their families to continue with traditional methods of cultivation. Allan has discussed instances where increasing density of population has led to a reduction in the length of the bush fallow, with consequent loss of soil fertility and a falling output of food crops.[2] Land-use systems and population densities are certainly related, but it is difficult to establish cause and effect, since systems of farming may themselves determine the critical density of population which can be supported in a given area without permanent damage to the environment. It could be argued equally well that the physical environment and the level of technology determine the population which can be supported in a given area. The most extensive systems of land use—hunting, food-gathering and nomadic stock-rearing—are found in arid regions and are associated with a sparse population, but it is difficult to see how an increase in population in these areas could cause a change to forest-fallow or bush-fallow methods of cultivation. Only a major capital investment in irrigation could make such areas capable of supporting a denser population.

In either case, from the point of view of practical farm management there is an important lesson to be learned, which is that extensive systems of farming are unlikely to be profitable in areas of high population density and small farms. The converse does not necessarily follow, that is to say intensive methods *may* be profitable in a sparsely populated area where there is plenty of unused land, although in theory this seems doubtful.

NATURAL RESOURCES

So far we have assumed that land is homogeneous, or all of one uniform type. This is clearly not the case; the total product which can be obtained per hectare and hence the population density which can be supported is influenced by the natural environment, the inherent fertility of the soil, the topography and climate. Generally in tropical Africa, water is the main factor limiting the distribution of vegetation and growth of plants, though the limits of crop production are probably determined by reliability, rather than by quantity of rainfall.

Thus in a sparsely populated region we would expect the more fertile land with more reliable rainfall to be cultivated first. Once all this land is in use, population growth will cause it to be culti-

[2] Allan, W. (1965) *The African husbandman*, London, Oliver and Boyd.

vated more intensively, but before this intensification has proceeded very far, labour may be used more productively by extending the area under cultivation onto land of lower fertility as shown

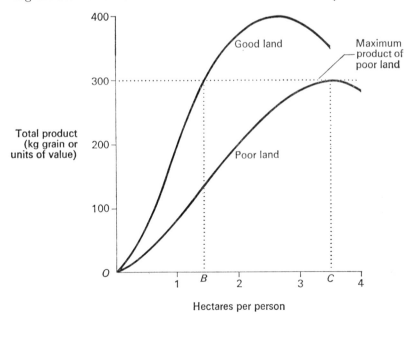

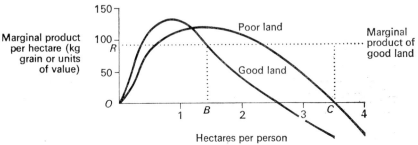

FIGURE 6.3 Extending production onto poorer land. Here we assume two types of land, good and poor. Once the area of good land per person falls to point *B* then the same total product per person can be earned by extending cultivation onto poor land. While the marginal product per hectare of good land is *OR*, the marginal product of poor land is zero.

in Figure 6.3. The total product per person is higher on a given area of good land than on the same area of poor land, so if there is plenty of good land available it would be irrational to cultivate the poor land. However, once the area of good land available per person falls to OB hectares, then the same total product per person can be earned by cultivating OC hectares of poor land.[3] At this stage labour is more productive if used to extend cultivation onto poor land rather than to intensify production on the good land. It should be noted that a larger area of poor land is needed to produce a given total product per person. Hence the poor land is farmed less intensively in larger units and supports a lower population density. So long as there is more poor land available, labour productivity is maximized by extending cultivation to the point where marginal product per hectare is zero as shown in Figure 6.3 at OC hectares of poor land per person. Yet because the good land is farmed more intensively at OB hectares per person it has a positive marginal product OR.

In fact, there is an almost continuous range of variation in soil fertility even between the plots of land cultivated by an individual farmer. In order to maximize profit from land of variable quality, farmers should use the principle already given in Chapter 4; labour and other inputs should be allocated between plots so that the marginal return is equal from all plots, as in Figure 6.4. These four plots are assumed to be equal in area but of varying fertility ranging from plot A the most fertile to plot D the least. If labour is allocated so that the marginal product per hour is the same on each plot, then plot A is farmed more intensively than plot B which in turn is farmed more intensively than plot C. In this hypothetical example, if the unit factor cost of labour is 1 kg grain per hour, then plot D should not be cultivated at all because the average product is less than the cost of labour, while it only just pays to cultivate plot C since the average product of labour just equals the cost of labour. Only a slight increase in labour cost, or a fall in productivity or product value, would cause plot C to go out of cultivation. It is therefore known as 'marginal land' since it would be the first to go out of production under adverse conditions.

Surveys of African farmers show that farmers do vary their labour and other inputs according to the fertility of different plots. Compounds are often established on the land with the highest natural fertility, so that the gardens round the house can be cultivated intensively with the aid of manures and household refuse; land further away which is also less fertile is then cultivated less intensively. In most areas there is land which is not worth culti-

[3] If rent was paid on the good land, then it would pay to extend onto poor land before the area of good land per person fell to OB hectares.

vating, even in densely populated regions, because it does not give an acceptable return to labour and other inputs. This is called sub-marginal land, while land which is more fertile than marginal land and therefore yields a surplus over the cost of labour and other variable inputs (or has a positive marginal product) is called super-marginal.

Agricultural production can be increased in two ways, either by intensifying production on land which is already farmed or by extending production onto land which previously was not farmed. If because of farming population growth or for some other reason the value of food products rises in relation to the value of labour, there will be a tendency for both intensification and extension to occur. In Figure 6.4, this new factor-product price ratio is

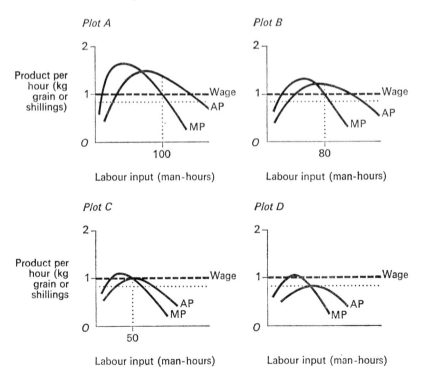

FIGURE 6.4 Land quality and intensity of use. Equal marginal returns per unit of labour are obtained if more labour is used per hectare on the high quality plot *A* than on the other plots of poorer quality, i.e. plot *A* is farmed more intensively. Plot *D* is sub-marginal and therefore not cultivated unless the wage per man-hour falls to 0·8 units (the dotted lines) when plot *D* becomes just marginal.

represented by the horizontal dotted line. The new economic optimum occurs at a higher level of labour input on plots *A*, *B* and *C* so production is intensified, but now plot *D* is no longer sub-marginal so production is extended onto this plot. Conversely, where agricultural growth and development is occurring the value of labour may rise in relation to the value of food products, and this may result in land which was previously farmed becoming sub-marginal. This may be seen on the desert fringes where land is going out of production. However these changes are usually associated with increasing intensity of capital use and hence increasing production per acre of the land remaining under cultivation.

Government-sponsored land settlement schemes are often established to extend production into areas of unused land, but generally the land is unused because it is sub-marginal under existing practices. New techniques and large amounts of capital may be necessary to make the land super-marginal and this may help to account for the high costs of some of these schemes.

Natural resources influence the crops which are grown and the livestock which are kept. Land which is sub-marginal for one product may be super-marginal for another; for instance good swamp-rice land may be quite unsuitable for cattle grazing. At the same time choice of products also influences intensity of farming. Intensive crops are those like vegetables that require large amounts of labour, fertilizer and frequently irrigation, or crops like rice and perennial tree crops which are grown continuously on the same piece of land without any intervening fallow. Extensive crops are those like pasture grasses and some cereals that, when not irrigated, require less labour per hectare and may be rotated with fallow periods. Nomadic cattle grazing is a particularly extensive livestock enterprise, whereas pigs and poultry fed on home grown or purchased cereals are intensive.

Generally, in tropical Africa, water is the main factor influencing the distribution of crops and livestock. Thus in the humid tropics two wet seasons occur each year, perennial tree cropping is possible and two crops of annual species can often be grown in a year, so intensive systems and high population density can be supported. In the drier, seasonally arid tropics the length of the dry season prevents the growth of perennial crops, which means that the seasonally arid land is sub-marginal for most perennial crops. Systems are less intensive. Areas which are drier still and where rainfall is unreliable may be sub-marginal for any kind of cropping, yet super-marginal for nomadic cattle grazing. There are no alternative uses for the land so for the graziers, the Masai of East Africa and the Fulani of West Africa, land has no opportunity cost; to them land is a free gift of nature.

Where water is less restricted there is greater freedom of choice of what to produce. The more fertile areas are super-marginal for many products; they have an absolute advantage over the poorer areas. However, this does not mean that production of all these products should be restricted to the more fertile areas. In fact everyone will be better off if each productive area concentrates on the enterprises for which it has the greatest *comparative* advantage.

Let us imagine just two farmers, Farmer A on good land and Farmer B on poor land, both producing just two crops, groundnuts and maize. For further simplification suppose labour to be the only variable input. The marginal products per day of labour for each crop on each farm are shown in Table 6.1. It will be noted that the

TABLE 6.1 Comparative advantage in groundnuts and maize production

	Marginal products per man-day in 50 kg bags	
	Bags of groundnuts	Bags of maize
Farmer A	$\frac{3}{4}$	1
Farmer B	$\frac{1}{4}$	$\frac{1}{2}$

marginal product of labour is higher for both crops for Farmer A so he has an absolute advantage over Farmer B for both crops. Clearly the total product would be increased if labour was transferred from Farm B to Farm A where the marginal product is higher. If this was not possible, total product could still be increased by transferring labour from one enterprise to the other on each farm and trading the products.

Thus suppose that Farmer A obtains 1 bag of maize from Farmer B. This means that Farmer A can spend 1 day less on maize production and still consume as much maize as before. If he uses that day in groundnut production he will produce an extra $\frac{3}{4}$ bag. Farmer B, in order to make good the 1 bag of maize sent to Farmer A, must transfer 2 days from groundnut production to maize production. His groundnut production will, therefore, be reduced by $2 \times \frac{1}{4} = \frac{1}{2}$ bag. In the circumstances if A gives B something between $\frac{1}{2}$ and $\frac{3}{4}$ of a bag of groundnuts in exchange for the 1 bag of maize both farmers will be better off. While both are using just the same amount of labour as before and both are maintaining their former levels of maize consumption, they each have more groundnuts than they had previously. It will be noted that these mutual benefits of increasing specialization and trading arise because Farmer *A* has a comparative advantage in the production of groundnuts, since his opportunity cost for $\frac{3}{4}$ bag of groundnuts is only 1 bag of maize whereas for Farmer *B* it is $3 \times \frac{1}{2} = 1\frac{1}{2}$ bags of maize. This means that Farmer *B* has a comparative advantage in maize production.

If we now apply suitable prices per bag to both groundnuts and maize, say £2 and £1·20 respectively, then we can see from Table 6.2 that this is simply a restatement of the principle of com-

TABLE 6.2 Marginal value products in groundnuts and maize production

	Groundnuts at £2 per bag	Maize at £1·20 per bag
Farmer *A*	£1·50	£1·20
Farmer *B*	£0·50	£0·60

parative advantage discussed in Chapter 4. Each farmer should transfer labour into the activity yielding the highest return. Each farmer will obtain the maximum return from each of his scarce resources when it is so allocated that the marginal value product per unit is the same in each enterprise. As we saw in Chapter 4, this may not lead to complete specialization but systems are likely to be more specialized than in areas of pure subsistence farming where there is no trade.

MARKETS AND COMMUNICATIONS

By its very nature, agriculture must remain widely dispersed over a large area, which means that a transport network is needed to take produce from the farms to the markets in the towns or places of export and to bring supplies and equipment to each farm. The location of farms in relation to the markets and the transport network has an important influence on farming systems. Thus in the most remote and isolated areas where there are no communications with the outside world, farmers are restricted to subsistence agriculture, but even where communications exist transport costs affect the intensity of farming and the kinds of crops and livestock produced.

The net return a farmer obtains from the sale of his produce is the market price minus the cost of getting the produce from his farm to the market, which in turn depends upon the distance involved. As a result it is profitable to farm land near the market more intensively than that further away. In fact location can have just as much influence on intensity of land use as the natural resources. Figure 6.3 may be used to illustrate the effect of distance from market, if 'good land' is taken to mean land near the market and 'poor land' is taken to mean land of similar fertility but yielding a smaller value product for a given level of labour use, because of higher transport costs. Thus land near the market can support a higher population density at a given level of money income per head than can more remote areas. This is clearly illustrated by the closely settled zone around Kano and other major African cities. Likewise Figure 6.4 could apply to plots of equal fertility at differing distances from the market. Then Plot *A* has the highest marginal product per hectare and is farmed most intensively because it is nearest the market; Plot *C* is marginal and Plot *D* sub-marginal because of their remoteness from the market.

Transport costs also influence the comparative advantage of different areas in the choice of what to produce since some products can be carried more cheaply than others in relation to their values. Cattle on the hoof cost less to transport than say fresh milk and vegetables, so remote areas have a comparative advantage in cattle production and areas around the towns a comparative advantage in milk and vegetable production. Generally speaking, transport costs vary with the weight, bulk, perishability and fragility of the products. Sugar cane, fresh fruits, timber for build-ing and firewood, and root crops like yams or cassava are relatively heavy per unit of value and are therefore costly to transport in comparison with cocoa, coffee, tobacco and rubber which weigh less in relation to their market price. Raw cotton is relatively costly to transport because of its bulk. Oil palm fruit is both bulky and perishable, since delay in extraction of the oil leads to a fall in its quality. Livestock carcasses, milk, fresh fruit and vegetables are also perishable; costs of transporting them are high, either because of refrigeration, or because of the high risk of physical deteriora-tion.

The relationship between transport costs and the location of production was studied in the early nineteenth century by a German economist von Thünen who developed the principle that products which have high transport costs in relation to their value will be produced nearer to consumption centres than products with low transport costs in relation to value. This means that products with low transport costs can be produced over a larger supply zone than can those more costly to move. Round any town with a demand for many foods and raw materials, the innermost zone is likely to supply the perishable products such as milk, vegetables, fruit and eggs; heavy and bulky root crops will also be grown fairly near the markets or along the major transport

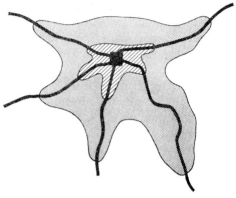

FIGURE 6.5 Von Thünen zones. The pattern of farming in the various zones might be as follows: inner zone—intensive continuous cropping of vegetables and perishable products; outer zone—arable with fallows, main food and cash crops; beyond outer zone—subsistence farming and cattle grazing.

routes; other products with low transport costs will be grown further out again. Such zones can be detected around many African towns and even small villages. However, these zones cannot be clearly distinguished in practice for various reasons. Other factors already discussed, such as the physical environment, also influence the pattern of cropping; the different zones of supply are unlikely to be circular since they extend along the main transport routes, as is shown in Figure 6.5, and overlapping of spheres of influence of adjacent markets create supply areas of irregular shape. Again, for various reasons it may be desirable to diversify into several different enterprises on each farm.

The third major factor affecting transport costs, besides distance and characteristics of the product, is the method of transport. Head loading is more costly than transport by pack animal or by animal-drawn carts, which are again more expensive than lorries.

Where water transport is possible, it is usually the cheapest method of all. Clearly, head loading is possible on the smallest of bush paths, whereas animal carts require broad tracks and lorries operate best on graded and metalled roads. Road and rail transport may have similar costs, although railways are generally cheaper for large consignments travelling long distances, whereas lorry transport is more flexible and therefore cheaper for shorter hauls and smaller consignments.

This means that the construction of a new road or railway leads to a considerable reduction in transport costs for all farms in the vicinity. As a result the supply zone for agricultural products is extended outwards, but at the same time, because of the saving in transport costs, all farmers in the area have an incentive to intensify production. Furthermore as communications and transport are improved, there is more opportunity for specialization of production in different regions, each tending to concentrate on those crops and livestock most suited to the physical environment. Thus the better the transport system, the more closely will the pattern of land use correspond to that best suited to the local variations in climate, soils and topography, since other products required for consumption can be bought from the markets.

The establishment of processing plants, such as sugar-cane crushers, cotton ginneries, palm-oil mills and fruit canneries in the main producing areas may also encourage increased production and specialization. In each case the processed product is less bulky, heavy or perishable than the raw material and as a result is cheaper to transport to the consuming centres.

These principles of location and transport have important applications to the organization and management of the farm apart from the overall influence on the farming system. The family home generally forms the centre of operations for the farm. The fields which are further away from the home incur higher costs of operation than do the nearer plots, on account of the greater amount of time spent travelling back and forth. This means that profits are maximized by cultivating plots near the home more intensively than those further away. Most African farmers do manage their farms in this way. As noted already, gardens around the house and compound are usually cultivated continuously with close interplanting and careful catch-cropping, fertility being maintained by manuring and composting. Outside this there may be plots of fairly continuous cropping maintained with some manuring but further away from the home the land is cultivated less intensively with longer fallow periods.[4] Similarly the pattern of cropping is influenced by location. For instance in a cocoa producing village in Ghana, Steel observed that the food farms were all within a mile or so of the town whereas the zone beyond was mainly under cocoa.[5] This is at least partly to be explained by the fact that food farms need to be visited at frequent intervals throughout the year while cocoa once established needs very little attention, no more than an occasional weeding between the trees and the plucking of pods when ripe.

In many parts of Africa, farmers have some freedom to choose the location and layout of their farms in relation to supplies of water, building materials and fuel and the markets or transport routes for their products. Clearly the objective should be to arrange the farm layout to minimize transport costs provided that total

TABLE 6.3 Examples of fragmentation

	Average size of holding (hectares)	Average number of plots per holding	Distance of furthest plot from homestead (kilometres)	Percentage of land that is fallow
Northern Nigeria[a]				
Hanwa	5·9	4·4	1·4	1·99
Doka				
(central village)	8·7	6·6	2·0	23·51
Dan Mahawayi				
(central village)	9·6	6·6	2·1	18·96
Kenya[b]				
Single extreme case				
of fragmentation	3·6	29	22·4	not available

Sources
[a]Norman, D. W. (1967) *An economic study of three villages in Zaria province: Part 1, Land and labour relationships*, Samaru, Nigeria, Institute for Agricultural Research, Samaru Miscellaneous Paper no. 19.
[b]Clayton, E. S. (1964) *Agrarian development in peasant economies: some lessons from Kenya*, Oxford, Pergamon Press.

[4] For example see de Schlippe, P. (1956) *Shifting cultivation in Africa: the Azande system of farming*, London, Routledge and Kegan Paul; Prothero, R. M. (1957) Land use at Soba, Zaria province, Northern Nigeria, *Economic Geography*, vol. 33, no. 1, p. 72; and Oluwasanmi, H. A., Dema, I. S. and others (1966) *Uboma—a socio-economic and nutritional survey of a rural community in Eastern Nigeria*, Geographical Publications Ltd., World Land Use Survey Occasional Paper no. 6.
[5] Steel, R. W. (1947) Ashanti survey, 1945–6: an experiment in social research, *Geographical Journal*, vol. 110, nos. 4–6, p. 159.

production is not affected. However, in many areas the land is shared out so that each farmer has some fertile land and some not so fertile, some land suitable for cash crops and some suitable for food crops. As a result farms are made up of many small scattered plots; they are fragmented. Examples of fragmentation are given in Table 6.3 for Nigeria and Kenya.

Fragmentation is wasteful of labour, since time is spent unproductively in travelling between plots; manure and fertilizers are seldom applied to the more distant ones. Mechanization, whether by animal draft or by tractor power, is only practicable on fairly large consolidated fields. Furthermore it is difficult to manage a fragmented holding as a single unit, since it is not possible to give proper supervision to labour and regular attention to crops and animals on all the plots when required. Consolidation of holdings and legal control of subdivision is considered by many authorities to be the essential base for the rapid development of commercial agriculture, whatever the system of land tenure.

LAND TENURE

Land has a social importance beyond its use as a productive resource, since the area controlled by any community represents the territory, the space for living, indeed the home of the individual members. Thus the nation of Kenya is associated with a particular area of land, so too is the Luo tribe. Every community claims certain territorial rights over a particular area and frequently opposes the alienation of land to outsiders. Indeed land is associated with religious traditions since it provides the burial place for ancestors and the heritage for future generations; as was stated by a Nigerian chief to the West African Lands Committee in 1912: 'Land belongs to a vast family of which many are dead, a few are living, and countless numbers are still unborn.'[6]

Today, there are wide disparities in the density of population, even within the boundaries of a single country, because of the general practice of restricting the occupation of land in each area to members of the dominant tribe. 'Tribal exclusiveness . . . is perhaps the most serious obstacle of all to the full development of the land', according to a publication of the Ministry of Animal Husbandry and Water Resources in Kenya.[7] It quotes the case

[6] Elias, T. O. (1962) *Nigerian land law and custom*, London, Routledge and Kegan Paul.
[7] Kenya Ministry of Animal Husbandry and Water Resources (1962) *African land development in Kenya*, Nairobi.

of the South Nyanza Luo who would neither allow families from other tribes into the Lambwe Valley scheme for new farms, nor settle in this underpopulated area themselves.

From the national point of view, the two main requirements of a system of land tenure are firstly, that it should lead to the most productive distribution of land among potential users and potential uses; and secondly, that it should provide sufficient security of tenure to justify measures to maintain or improve the productivity of farms. To some extent these two requirements are in conflict, since the greater the security of tenure, the more difficult it becomes for efficient farmers to obtain control of land occupied by the less efficient. Most of the customary tenures in Africa also emphasize a third feature: the right of each family to some share in the natural resources belonging to the community, which may be an extended family, a clan, a village, or a tribe of several thousand or even million people. Communities do not normally permit their members to dispose of land by way of sale or lease to strangers, but, subject to these overriding controls, individuals belonging to the community have the right either to use particular pieces of land which may have been inherited, or to occupy plots under shifting cultivation. Members of pastoral tribes have similar rights to graze cattle over the territory recognized as belonging to the group in general, with all the associated rights of using water, and perhaps growing a little grain.

African patterns of land tenure have proved to be flexible in allowing for differences in the needs of families and for changing circumstances. It has generally been possible for the more energetic or capable farmers to obtain extra land, either from that controlled by the community and still unused, or by way of pledge or loan from other families, who have more land than they can use. Again, the introduction of tree crops in both East and West Africa led to an extension of individual rights in land to cover the prolonged period of occupation and to allow the pledging and even sale of cocoa or coffee plots which a man has built up by his own work. In such cases systems of land tenure may evolve naturally through increasing individualization to the stage of a commercial market in land, where plots can be bought and sold like other commodities. However, the existing pattern of land tenure may not evolve quickly enough to avoid acting as a constraint upon agricultural changes which are urgently required in view of the rising populations and the expansion of export crop production. In the first place the loan or pledging of land, based on close personal relationships, provides neither the security of tenure nor the incentive for efficient farming by the occupiers. Secondly, the sale of land becomes associated with lengthy and costly legal disputes regarding the true ownership, particularly if it is sold to strangers from outside the community. Hence, with the development of a market in land, it becomes important to establish individual rights on a legal basis by survey and registration of ownership. Where the individual owner has the sole right to use a piece of land or to dispose of it as, and to whom, he wishes it is known as freehold tenure. Similarly at least in arid areas, it may be necessary to establish and register rights to draw water for irrigation or other purposes from a particular source.

It is widely believed, particularly in East Africa, that individual freehold tenure is a highly desirable, even an essential, component of any agricultural development programme. The pride of ownership and the security offered to the farm family by this form of tenure are thought to encourage long-term improvement and conservation of the land and associated water resources. Furthermore privately owned land can be offered as security for loans, to be given up to the lender in the event of failure to repay the loan; that is it can be mortgaged, thus enabling small farmers to raise money for farm improvement. However, this argument is of doubtful validity since few banks and commercial moneylenders are prepared to accept land as security for loans because of the practical and political difficulties of removing farmers from their land if they fail to repay their loans. It is also argued that the market in freehold land encourages the able and industrious farmers to expand production by buying the land of the less successful thus encouraging the development of a commercial attitude to farming. Individual freehold tenure is well established in Buganda under the *mailo* land system, where some owners farm their own land and others lease out some or all of it for an annual rent.[8] In other parts of Uganda and in Kenya official policy is directed towards the establishment and registration of individual freehold rights.

In contrast, many socialists and others find freehold tenure unacceptable since they believe that land, water and capital should belong to the community rather than to the individual. They argue that private property in land can lead to class distinctions between landless labourers and landowners who may be

---

[8] See West, H. W. (1964) *The* mailo *system in Buganda: a preliminary case in African land tenure*, Entebbe, Government Printer.

able to live on the returns from land ownership, without needing to work. Thus the landowner may employ labourers to work his land for wages, while he takes the surplus production for his own consumption. At the same time land may gain in value as population density increases, as new techniques improve the productivity of the land or as new roads and markets open up the area, so profits could be made simply by buying and selling land. Alternatively, landowners may lease the land to tenant farmers, in return for rent payments which some would view as unearned income. Increases in population reduce the marginal product of labour and raise the marginal product of land so that the tenants' average incomes fall while the wealth of the landowner rises. This may eventually lead to political unrest. State control of land already exists in some African countries, where the former colonial governments assumed rights of ownership of so-called 'crown lands' as in Northern Nigeria, Botswana and Swaziland. These lands are now controlled as 'state lands' by the independent governments of these countries who can, if they wish, exercise rights of ownership. Large farms can be established under state direction, or the land can be leased out to tenant farmers in return for rents.

Leaseholds exist in a variety of forms. Under customary tenure, land may be borrowed in return for annual gifts which acknowledge the prior rights of the lender; they are often not related to the returns to be earned from the land, and so do not correspond to the economist's idea of rent. More formal systems of leaseholds require annual payments of rent for the use of land on agreed contractual terms and for a specified period. Under conditions of pure competition in the market for leasehold land, where there are many alternative users competing among themselves for the hire of land, rent represents the surplus that a particular piece of land is expected to produce, over and above the cost of all other inputs including the living expenses of the farm families. The rent or surplus per hectare of land is, in fact, the marginal product per hectare as shown in Figures 6.2 and 6.3, which differs for different pieces of land. Competitive rents must increase with increasing population density, with increasing fertility of the land and with increasing proximity to a main market or transport route. Marginal land has zero marginal product and hence cannot command a rent.

The payment of competitive rents ensures the most productive distribution of land among potential users and potential uses. Thus if a particularly able or energetic farmer can produce a higher marginal and average product per hectare of a particular piece of land than his neighbours, he can outbid them in the rent he is willing to pay. In this way the better manager may control more land than his less able neighbour. Furthermore farmers are obliged to operate the most profitable system of farming that the land is capable of supporting, in order to afford the rent payments. Thus it is unlikely that a system of shifting cultivation would yield sufficient output to meet all the costs of production, including a competitive rent, on land which is wanted for continuous vegetable growing.

The annual rent payments are fixed by contractual agreement between the tenant and the landlord who may be a private land-owner or a representative of the state where the land is state owned. The payments may take the form of a fixed sum of money or a fixed share of the main crop output. Payment in cash has the advantage of leaving tenants free to make their own choice of what crops to grow on the basis of their relative profitability; the annual payment of rent becomes part of the fixed costs of each farm, which must be met from the farmers' incomes if they are to continue in business. Where the main crop is shared, farmers are restricted to growing that crop unless new agreements can be negotiated with the landlord. This is not a serious problem in the case of a permanent crop or where one crop is clearly more profitable than any others. Thus, on the Gezira scheme in the Sudan, crop sharing was satisfactory so long as there was a clear advantage in cotton growing, but now some dissatisfaction is arising since some tenants would like to grow more of other crops.[9]

Since they are based on the expected productivity of land over a period of years, cash rents do not usually allow for a run of exceptionally bad seasons, for which the tenant must take all the risk. On the other hand, the sharing of annual crop yield between landlord and tenant means that risks of crop failure are also shared. This is a particular advantage where the state, as land-lord, is trying to encourage the introduction of a new crop or technique such as irrigation. The major criticism of crop sharing arrangements is that they diminish the inducement for the tenant to increase production since the landlord shares in the benefits but not in the variable cost of production. This is illustrated in Figure 6.6, where the thick curve represents the expected marginal product per unit of labour or other variable input and the thin

---

[9] Thornton, D. S. (1966) *Contrasting policies in irrigation development: Sudan and India*, University of Reading, Department of Agricultural Economics, Development Study no. 1.

curve the net return, assuming that the landowner takes half the product. The maximum profit for both parties occurs at E_2 but under this system of share-cropping the tenant maximizes his profit at E_1.

The productive use of natural resources depends not only on the level of variable inputs but also on the application of capital for conservation and improvement. Certain capital costs are incurred in preparing land for use, in clearing, draining, fencing and providing buildings, access roads and paths but further capital inputs in irrigation works or establishing tree crops can increase productivity. The capital may be provided by the landlord or the tenant.

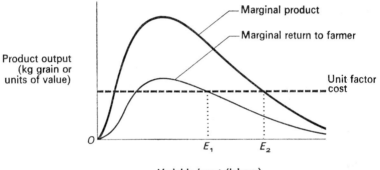

FIGURE 6.6 Effect of share-cropping on variable resource inputs. The thick curve represents marginal product per unit of labour; the thin curve the net return if the landowner takes half the product. Under other systems of land tenure the economic optimum would occur at E_2 but under share-cropping it is reduced to E_1.

Where it is provided by the landlord the contractual rent payments include a return on the landlord's capital, indeed this may be the major part of the rent paid. However, some of the capital must inevitably be provided by the tenant, particularly that involved in conserving and maintaining existing resources, in preventing soil erosion, in maintaining paths, irrigation canals and established tree crops. Farmers will not usually devote labour and money to such tasks, unless they will reap the benefit by the continued use of the particular piece of land which they have maintained or improved, but leaseholds between private individuals may provide little or no security of tenure. This is illustrated in the rubber producing areas of Mid-Western state of Nigeria where the trees are tapped by share-croppers who come from other parts of the country. The too frequent and too careless tapping leading to the early death of the trees may be blamed, at least in part, on the insecurity of tenure of the share-croppers. Government intervention may be needed in such circumstances to ensure that tenants do not lose their farms except in extreme cases of bad farming. On government-run settlement and irrigation schemes, farmers are normally given security of tenure provided that they occupy their farms with their families, cultivate to approved standards and pay their rents and other dues regularly. They are usually prohibited from dividing their farms and from subletting but tenants may bequeath their holding to one approved heir.

Alternatively, on other settlement schemes, settlers are expected to buy the freehold rights to their farms with the help of a mortgage. When a farmer buys land he is paying for a stream of future incomes or rents and the price represents the present value of this stream of incomes. Thus although land is costless, where there is competition for its use it has a price as a consequence of its marginal product or rent. Having bought a piece of land, a farmer has an incentive to earn a satisfactory annual rate of return or interest on the money he has spent. However, once he owns the land and has paid off any mortgage then he has no further annual payments to make for the use of the land. To that extent, he *may* farm less intensively than a tenant who must pay an annual rent. Nevertheless, under any system of tenure, including traditional customary tenure, if there is competition for the use of land, this resource will have an opportunity cost which must be taken into account in rational decision making, even though no payment is made.

In summary, it is doubtful whether land reform, meaning a government policy aimed at changing patterns of land tenure, is needed in much of Africa. The argument that customary tenure offers no security or inducement for conservation and permanent improvements to the land has proved unfounded in many cases. Although customary tenure does not permit land to be offered as security for loans, it is questionable whether mortgages are a satisfactory form of credit even where farmers own their land. Government intervention may be needed for the consolidation of plots and prevention of further subdivision where farms are excessively fragmented but it should be noted that if not controlled fragmentation is just as likely to occur under freehold as under communal tenure. Where the buying and selling of land has become accepted, registration of rights to land and water may be desirable to reduce the opportunities for disputes and ill-feeling between neighbours. The establishment of individual rights to sell land even to strangers from outside the local community or direct government intervention to nationalize land resources may be necessary to break down tribal barriers and to enable population movements from more densely populated regions of the country to those which are less densely populated.

Apart from the planning of towns and of land for industrial uses, many cases arise in rural areas where common action is required over a large area of land in the interests of a community as a whole. Irrigation may require the building of dams and canals, the prevention of erosion may require the planting of forests or the control of grazing and cultivation; so the community or the state must acquire control over a large area of land. This requires prolonged negotiation with the individuals concerned whether they hold their land under customary tenure or freehold title. The human and social aspects of land consolidation and improvement schemes probably require more careful preparation and planning than the technical aspects.

Communal grazing raises its own special problems since under this system it is difficult to improve livestock or grassland. In the first place it is hard to keep cattle from straying too far and even from trampling on adjacent cultivated land, so many people spend their days herding livestock and trying to keep them where they belong, seldom with complete success. Second, many common lands, particularly in the region of watering points, are overgrazed. The grass is kept too short to grow well and the better quality grasses die out entirely; eventually the grass cover may be destroyed and erosion results. Because the land is free to all, no one has an individual incentive to limit the number of animals he turns on to it, to attempt to maintain or improve the grazing or to provide more watering facilities. Finally, progressive farmers who want to improve their own animals cannot do so as long as their stock graze together with that of others on the common land. They cannot control diseases, nor can they control breeding so as to improve their herds.

Clearly enclosure and fencing is required but this is generally opposed by the livestock owners and it raises difficulties where seasonal migration is necessary to find water and grazing. However, it is the enclosure of individually owned grazing land which is most strongly opposed. In such circumstances, a better alternative for governments concerned with the improvement of livestock management may be the encouragement of cooperative groups to deal with cattle management. The whole group must then be persuaded to agree to fencing the grazing land, limiting stock numbers and improving breeding and disease control.

SUGGESTIONS FOR FURTHER READING

ALLEN, W. (1965) *The African husbandman*, London, Oliver and Boyd.
BOSERUP, E. (1965) *The conditions of agricultural growth*, London, George Allen and Unwin.
BROCK, B. (1969) Customary land tenure, 'individualization' and agricultural development in Uganda, *East African Journal of Rural Development*, vol. 2, no. 2, p. 1.
CHAMBERS, R. (1969) *Settlement schemes in tropical Africa, a study of organization and development*, London, Routledge and Kegan Paul.
CHISHOLM, M. (1968) *Rural settlement and land use*, 2nd ed., London, Hutchinson.
DUNN, E. S., Jr. (1954) *The location of agricultural production*, University of Florida Press.
JOHNSON, O. E. G. (1970) A note on the economics of fragmentation, *Nigerian Journal of Economics and Social Studies*, vol. 12, no. 2, p. 175.
OLUWASANMI, H. A. (1957) Land tenure and agricultural improvement in tropical Africa, *Journal of Farm Economics*, vol. 39, no. 3, p. 731.
PEDRAZA, G. J. W. (1956) Land consolidation in the Kikuyu area of Kenya, *Journal of African Administration*, vol. 8, no. 2, p. 82.
RUTHENBERG, H. (1971) *Farming systems in the tropics*, Oxford, Clarendon Press.
UCHENDU, V. C. (1967) Some issues in African land tenure, *Tropical Agriculture*, Trinidad, vol. 44, no. 2, p. 91.
WHETHAM, E. H. (1968) *Co-operation, land reform and land settlement*, London, the Plunkett Foundation for Cooperative Studies.

7 Labour

THE MEASUREMENT OF LABOUR

The resource called labour is the work done by human beings and not the persons themselves. It should therefore be measured as a flow over a given period of time. If we say that a particular farm has a labour force of 3 men we imply that 3 men are employed full-time throughout the year. If 1 of the 3 had another part-time job or travelled away for six months of the year we might say that the labour force was between 2 and 3 men, say $2\frac{1}{2}$ men.

A distinction should be made between the amount of labour available and the amount actually used. The two may differ where there is unemployment or underemployment of labour, either permanent or seasonal. The definition of the amount of labour available is somewhat arbitrary, depending as it does on who is included in the labour force and how many hours they are willing and able to work. The size of the family labour force depends upon the age at which children are expected to help on the farm or in other productive activity, and whether women and old men are included. Similarly the hours available per person per year depend upon the number of hours individuals are prepared to work and the extent of off-farm commitments such as housework and trading for wives and school attendance for children. Thus the size of the labour force and the hours worked depend upon customs and tradition and attitudes towards knowledge, leisure and income. There is really no justification for assuming that standards accepted in other parts of the world, such as the 48-hour week, should apply to African farmers. However, these social characteristics are themselves influenced by the opportunities for economic gain, as measured by the marginal productivity of labour. The influence of the marginal product on the supply of labour will be discussed later in the chapter, as will the seasonal underemployment which is inevitable in agriculture.

The amount of labour used, that is the actual labour input, over a given period on a particular farm or plot of land, depends upon the number of individuals employed, the number of hours they work and their rate of working per hour. In principle, there is no difficulty in measuring labour inputs in terms of numbers employed and hours worked, although in practice farmers may not have clocks and estimates of the passage of time based on the movement of the sun may be imprecise. Rates of working vary according to the task, the crop, the cropping sequence, the soil type and condition, the plot size, the tools and methods used and the sex, age, nutrition and health of the workers. Most farmers know approximately how long a given task will take from past experience but for new methods and machines the measurement of rates of working requires careful recording of the time taken and the amount of work done. In fact 'work measurement' is one aspect of the specialist field of 'work study'. Measurement of the amount of work done may raise difficulties. For cultivation tasks the amount of work may be assessed in terms of the area covered but for weeding or harvesting the amount of work involved is affected by the weed growth or the crop yield as well as the area. Processing tasks may be measured in terms of the quantity of produce handled but it may be more difficult to define the amount of work involved in caring for livestock. In every case the quality of work may vary and this should be taken into account in assessing the work done.

For many farm management purposes, variations in the rate of working are ignored and labour inputs are measured in man-hours. The labour input in man-hours is the product of the number of men employed and the average hours worked by each. It is therefore assumed that the labour of 1 man for 100 hours is equivalent to the labour of 100 men for 1 hour or 5 men for 20 hours. This assumption is slightly unrealistic, but it is adequate for most purposes, provided it is realized that there is a time limit on most tasks, so that more than 1 man may be needed to finish within the time available. However, where women and children are employed on farm work, their rates of working may differ from those for men so that a woman-hour or a child-hour is not equivalent to a man-hour. In fact the relative rates of working for men, women and children will vary from one task to another. Men may cultivate faster but harvest or weed more slowly than women. To some extent the division of labour and specialization of different age groups and sexes on particular tasks may reflect these differences in rates of working. Thus women tend to specialize in crop harvesting, processing and marketing and the care of small livestock. Children help with planting, weeding and harvesting. Hired labourers may be employed on heavier tasks such as bush clearing and the care of cash crops.[1]

For these reasons, it may be desirable to treat the labour of men, women and children as separate and distinct inputs for purposes of farm analysis and planning. However, this is inconvenient, particularly where a task is done by the whole family group. The alternative is to use some system of weighting to convert hours worked by women and children into man-hour equivalents. However, when a woman works half the speed of a man on one task and twice as fast on another, fixing her equivalent value at 0·5 on the basis of the first will grossly underestimate family labour capacity on the second. There are certain tasks which occur at busy times of the year and the relative rates of working on these critical tasks should form the basis for establishing man-hour equivalents. At present relatively little precise information is available on relative rates of working so conversion to man-hour equivalents is rather arbitrary. One scale of conversion commonly used is based on the assumption that the work done per hour by women is two-thirds, and that by children under 15 years of age, one-third of that done per hour by men. The number of hours worked by women is multiplied by two-thirds and the number of hours worked by children is multiplied by one-third before adding these to the hours worked by adult males to give the total labour input in man-hour equivalents.

THE PRODUCTIVITY OF FAMILY LABOUR

The unit of agricultural production in Africa is normally the nuclear family of one man, his wife or wives and children; there may be other dependents such as an elderly parent, younger brothers or children of relatives. This basic unit may, however, contain a number of separate economic units since each wife and grown son may cultivate plots for their own use, in addition to providing labour for the main farm, the output of which is controlled by the senior male.

Traditionally, this nuclear labour force was supplemented on occasions by kinsmen or members of the farmer's age group brought in to help with particularly heavy work, such as hut building or clearing land. The head of the family paid for this labour, partly

---

[1] See Upton, M. (1967) *Agriculture in south-western Nigeria*, University of Reading, Department of Agricultural Economics, Development Study no. 3.

in providing food and drink and perhaps music for dancing at the end of the task; and partly in accepting an obligation to work in a similar manner for those who had rendered him assistance. Such a system of neighbourly cooperation added to the social life of the extended family or the village, rather than to the total supply of labour.[2]

Thus the labour force was initially restricted to the nuclear family and the size of the farm depended upon the number of active family members. So long as the marginal product per person was greater than the cost of subsistence a large family was desirable, among other reasons, because a man could thereby increase his wealth and power.

However, as shown in earlier chapters, after a point the addition of more workers to a fixed amount of land and capital will bring diminishing marginal and average products per person, with given farming techniques. Eventually if there is no alternative productive employment other than farming, population growth on a fixed area of land may drive the marginal product per person down to the subsistence minimum. Beyond this point the marginal products of any additions to total population would be inadequate to keep them alive. Poverty and undernutrition would keep the population constant at the subsistence level of income.

It is possible that under the system of social security provided by the extended family, labour may be supported *beyond* the point where the marginal product falls to the subsistence minimum. This is illustrated in Figure 7.1, which shows hypothetical total, marginal and average product curves for increasing labour use on a given area of land. According to this graph the marginal product of the third person employed is greater than his subsistence needs but that of the fourth person is not; the economic optimum occurs at point *A*, just beyond 3 persons. The cost of supporting the fourth is greater than the increase in production. However, there is a surplus over and above the subsistence needs of the rest of the family, shown by the shaded area; remember the total product is the sum of the marginal products so the total product is represented by the total area under the marginal product curve. The total subsistence need of the family is the subsistence need per person times the number of persons, so it is the area under the horizontal line. If some of this surplus is redistributed among the family

[2] See Alao, J. A. (1967) Reciprocity—mutual aid systems, *Bulletin of Rural Economics and Sociology*, Ibadan, vol. 2, no. 2, p. 114.

members, the fourth person may receive enough produce to sustain him and possibly the fifth.

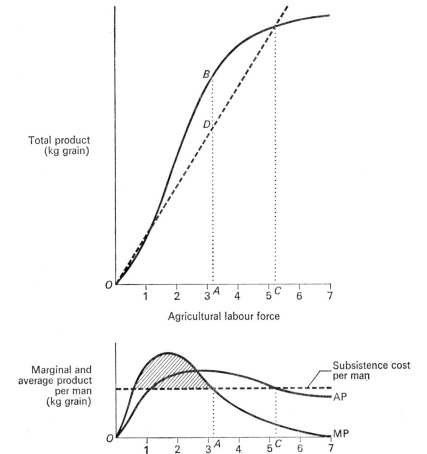

FIGURE 7.1 Disguised unemployment. In the upper diagram the distance *BD*, and in the lower diagram the shaded area, represents profit, which may be redistributed among family members. If this is the case the maximum number of people that can be supported is *OC* where average food cost equals average food product per person. The marginal value product is well below the unit factor cost at this point.

In fact if the whole total product is available for family consumption, the average product is a better measure of the amount available per person than is the marginal product. Then the maximum number which could be supported on the family farm is given at point C where the average product has fallen to the subsistence minimum. By this stage the marginal product of labour may have fallen even to zero. A situation like this where all the family are apparently employed on the farm but where the marginal product is less than the subsistence wage (the unit factor cost of labour) is known as 'disguised unemployment'. If it exists it means that labour can be withdrawn from farming without reducing the marketable surplus, in fact the withdrawal of labour might increase it. The labourers withdrawn from agriculture were adding more to total costs than to total farm output.

Disguised unemployment may or may not exist in parts of Africa. There is no clear evidence. It is difficult to measure the marginal product of labour, because of seasonal variations and because many farmers and their families have minor secondary occupations. However, on theoretical grounds it may be argued that the scope for disguised unemployment is limited. This is because the cost of other factors of production besides labour must be met out of the total farm product. In Figure 7.1 the quantity BD or the shaded area is really the marginal product of land, capital and other fixed resources. Under a landlord-tenant system the bulk of this surplus may go as rent. Even if no rent is paid, some items of capital are necessary for any kind of farm production and some of the surplus must be set aside to provide for these capital needs. Whether or not there is disguised unemployment, there are without doubt some areas of Africa where the return to labour is virtually at the subsistence minimum. Population growth is limited by the supply of food and agriculture cannot be improved without the introduction of capital and new techniques from outside.

Increases in the family labour force need not be accompanied by diminishing marginal returns per person if other inputs are increased to keep pace or if new techniques are introduced. Thus when all inputs of land, capital and management are increased at the same rate as labour inputs and when there are constant returns to scale, marginal and average products per man remain constant. Alternatively if the area of land is fixed it may be necessary to increase capital inputs faster than labour inputs in order to maintain a constant rate of return to labour. The introduction of new techniques may enable the marginal and average products per unit of all resources to be maintained or even increased. However most new techniques involve the use of new capital in the form of machines and equipment, stocks of new varieties of seeds, and agricultural chemicals or new breeds of livestock and manufactured feedstuffs. Thus we say the new techniques are 'embodied' in the new forms of capital.

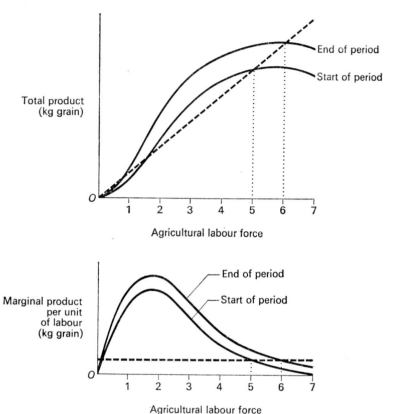

FIGURE 7.2 Growth in population and output. If because of technical change and capital investment the response curve for labour is shifted upward, then population growth need not result in diminishing marginal returns. Thus if the population is 5 units at the start of the period and less than 6 units at the end of the period then marginal and average products per unit of labour will increase.

Over a period of time, say ten years, the combined result of increasing capital inputs and the introduction of new techniques will be an upward shift in the production response curve for labour as shown, using hypothetical data, in Figure 7.2. Now according to this diagram, if the family labour force increases by 20 per cent, that is from 5 persons to 6 persons, over the ten-year period the marginal and average product per person will be the same at the end of the period as at the outset.[3]

If the labour force grows by less than 20 per cent there will be an increase in the marginal and average product per person and conversely if the labour force grows by more than 20 per cent the marginal and average product per person will fall. These relationships are also illustrated in Table 7.1 where the left-hand column

TABLE 7.1 Changes in labour force and total farm product: hypothetical figures

Percentage increase in labour force over ten-year period	Corresponding percentage increase in total farm product
0	10·0
5	12·5
10	15·0
15	17·5
20	20·0
25	22·5
30	25·0

After Tuck, R. H. (1960) *An introduction to the principles of agricultural economics,* London, Longmans, ch. 12; and Tuck, R. H. (1968) The choice of objectives in economic development, *Mediterranea,* Paris, no. 17, p. 6.

shows possible alternative rates of change in the labour force and the right-hand column shows the corresponding increases in the total output of grain. The main conclusion to be drawn is that if the labour force is growing, some increases in the use of capital and new techniques are necessary, just to keep pace, that is to avoid diminishing marginal and average returns per person employed. More rapid growth of capital inputs and innovations will bring about increasing marginal and average products per unit of labour and hence increasing incomes for those engaged in farming.

HIRED LABOUR AND OPPORTUNITY COSTS

So far we have assumed that the family is the sole source of labour and there is no alternative employment. In many parts of Africa hired labour is used; by the coffee farmers of Uganda and Tanzania,

[3] We are making the simplifying assumption that the relationship between marginal and average product (i.e. the elasticity of response) remains constant. If we relax this assumption the rate of growth of the labour force at which marginal product remains constant will differ from the rate of growth at which average product remains constant.

the cocoa growers of West Africa and commercial farmers in Kenya and Northern Nigeria for instance. Labour is then a purchased input, the wage rate representing the unit factor cost. The economic optimum level of labour hire occurs where the marginal value product is equal to the wage rate per unit.

Labour may be hired on a regular basis possibly for a period of several years, on a seasonal basis to help with harvest or some other major recurring task, or on a casual basis. Casual hired labour is likely to be provided by local farmers or other villagers in an emergency or for a special purpose. Many farmers prefer to hire labour on such occasions since the cost of feeding and entertaining 'voluntary' helpers may exceed the wages of hired hands. Regular and seasonally hired labourers are frequently migrants from poorer regions. They may live with the employer's family and share his meals or they may be allowed to establish food crop farms of their own. The food they receive or the free use of land represents a part of their wage payment. For permanent labour and some seasonally hired labour it is not possible to adjust the level of employment to precisely the point where marginal value product equals the wage rate, since men are not divisible. However, the farmer has found his economic optimum level of labour employment if: (1) it is not possible by reducing his employment of labour to save on costs more than he would lose in receipts; (2) it is not possible by increasing his employment of labour to add more to his receipts than to his costs.

The expansion of the exchange economy creates new economic opportunities for men and women outside the farm. Thus adult males may migrate to mines, plantations or towns in search of wage employment. Even without migrating, farmers and their wives may find secondary occupations in the village. In particular food processing and trading in imported foods, matches, candles, clothing and so on are frequently found. In addition men may find employment in blacksmithing, carpentry and other crafts. Education is widely viewed as a means whereby young men may leave farming and earn higher incomes elsewhere. Not only are children kept in school instead of working with their parents on the farms, but those leaving school are not prepared to return to farm work. Generally their parents approve of this attitude. Over half a sample of 153 farmers in south-western Nigeria said they would prefer their children to take non-farming jobs and to live in towns rather than to have them working on family farms.[4]

4 Upton, *Agriculture in south-western Nigeria*.

Where there are alternative employment opportunities, the highest wage rate which could be earned represents the opportunity cost of employing a family member on the farm. If the value of his marginal product is less than this opportunity cost total family income would be increased if he left the farm and worked for wages. However, in most African countries there is severe urban unemployment. Many of the young men who move to town hoping to earn high wages are unable to find work. Both from the point of view of individual families and in the national interest it is desirable that many of these young men should find work in agriculture. However, as an incentive it is necessary to raise farm incomes by increasing capital inputs and introducing new techniques of farming.

HOURS WORKED AND LEISURE PREFERENCE

It is sometimes argued that economic analysis does not apply to peasant cultivators because they do not respond to economic incentives in that they do not always work harder if the marginal product or wage per hour is increased. This of course is nonsense. In every society in the world people value their leisure and will not work all the hours that are physically possible in order to earn more. In fact every hour of work has a subjective cost to the individual. He will only work so long as he values the product of his effort more than he values his leisure. If an individual's income per hour of work increases, the opportunity cost of leisure is increased. Hence he is tempted to work more hours; that is to substitute work for leisure. Where this is the case the relationship between income or return per hour of labour and the hours worked, known as the supply curve for labour, will be as shown in Figure 7.3.

However, it may be that a farmer wants only the basic minimum of food, clothing and shelter for his family; once he has these, he prefers leisure to further effort. In such cases, if the average product per hour of work rises, the family will work fewer hours. Conversely, if the average product per hour falls, the farm family must work more hours to obtain the so-called 'target income'. The supply curve is then a perverse or 'backward-sloping supply curve' as shown in Figure 7.4. A preference for leisure does not in any way imply irrationality. Thus it is said that the African subsistence farmer's preference for shifting rather than sedentary cultivation is really one of preference for leisure; but if the family's subsistence needs can be met with less effort under shifting cultivation, this is the more rational system of farming to pursue. If, however, because of population pressure, more intensive methods become necessary and as a consequence the average product of labour falls, farmers and their families must work harder and longer hours to provide for their subsistence needs. Their leisure time is reduced.

Furthermore, when we speak of leisure we are really referring to time spent in non-agricultural pursuits. We must remember

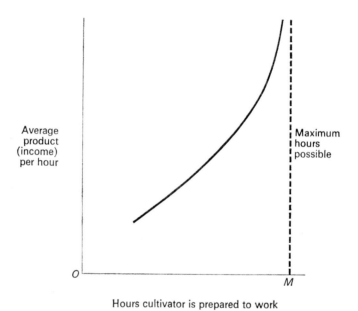

FIGURE 7.3 Normal supply curve of effort. Increases in income per hour encourage men to work longer hours.

that out of their total of non-agricultural hours, subsistence producers have to find time to make their clothes, build and repair their houses, perform numerous religious and civic duties and many other tasks which commercial farmers have done for them. In an area of subsistence agriculture, the less time farmers spend on their farms the more time they have to build bigger and better houses or make finer clothes. An increase in the average product per hour of farm labour will clearly give an incentive to work less hours on the farm. The labour supply curve will slope backwards. This will not be the case if cultivators are undernourished. Then an increase in the average product per hour will provide them with the necessary energy to work longer hours. The supply curve will slope forward.

Once there are opportunities for earning and spending cash, the cultivator is likely to revise his ideas about leisure; or more specifically he is likely to revise his target income upwards. On top of his subsistence needs he will require income to pay taxes or school fees and to buy consumer goods like bicycles, radios, furniture and new clothes, building materials and foods not produced locally. Then the hours cultivators are prepared to work may remain constant or even increase as the average return per hour of labour increases.

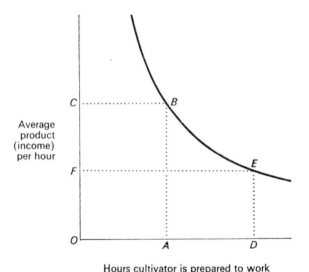

Hours cultivator is prepared to work

FIGURE 7.4 Backward-sloping supply curve of effort. Area $OABC$ = area $ODEF$ = target income. If there is a target income, increases in income per hour mean that it can be earned by working fewer hours.

Ultimately the total number of hours cultivators can work is fixed by the absolute limitation on the hours of daylight or by biological need for rest. So the maximum possible labour supply per person may be 12 hours per day. As the hours worked approach this level the subjective cost of forgoing leisure will be greater. This means that the supply curve will rise more steeply as shown in Figure 7.3. If average return per hour continues to rise the farmer may prefer to take some of his increased prosperity in the form of leisure. Once again the supply curve for labour may be backward sloping. This phenomenon of leisure preference is not confined to peasant farming areas. In any society, there will be some level of hours worked beyond which a man will prefer additional leisure to additional cash income.

It has even been argued that the perverse labour supply curve of work may mean that farmers who have reached a certain level of prosperity produce less as their income increases when for instance the prices of their products increase. This is most unlikely. Even if the individual's labour supply curve is backward sloping an increase in the relative price of agricultural products will attract other resources, such as capital, into agriculture and out of other industries. It may even attract workers out of other industries into agriculture so that the total labour supply may be increased even if individuals work less hours.

Similarly an increase in the price of a single product, say cotton, will attract labour and other resources away from the production of other crops. In fact various studies have shown that the quantity of a specific crop grown increases when the price increases.[5]

The backward-sloping supply curve has been observed for temporary migrant labour and is to be expected in this case.[6] If a man requires a particular sum of money in order to pay taxes, buy a bicycle or accumulate a bride price he has a very clear target income. Furthermore the subjective cost of being away from his home and family is very high and likely to increase rapidly the longer he is away. The backward-sloping supply curve is less likely for labour inputs on a man's own farm.

It is therefore impossible to generalize about the supply response of effort to an increase in the average return. Indeed the response is likely to vary from one individual to another. Nevertheless when compared with agricultural workers in temperate regions and with commercial and industrial workers in Africa, African cultivators work relatively few hours on their farms. Thus a typical wage labourer elsewhere might work about 8 hours per day for 250 days a

[5] See Ady, P. (1968) Supply functions in tropical agriculture, *Bulletin of the Oxford Institute of Economics and Statistics*, vol. 30, no. 2, p. 157; Bateman, M. J. (1965) Aggregate and regional supply functions for Ghanaian cocoa, 1946–62, *Journal of Farm Economics*, vol. 47, no. 2, p. 384; Dean, E. (1966) *The supply response of African farmers: theory and measurement in Malawi*, Amsterdam, North Holland; Maitha, J. K. (1969) A supply function for Kenyan coffee, *Eastern Africa Economic Review*, vol. 1, no. 1, p. 63; Stern, R. M. (1965) The determinants of cocoa supply in West Africa, *in* Stewart, I. G. and Ord, H. W., eds., *African primary products and international trade*, Edinburgh University Press.

[6] See Berg, E. J. (1961) Backward-sloping labour supply functions in dual economies —the Africa case; *Quarterly Journal of Economics*, vol. 75, no. 1, p. 468.

year, giving a total of 2000 hours per year. Some estimates for African cultivators suggest fewer hours worked per day and fewer days worked per year (see Table 7.2). This difference does not necessarily mean that the cultivators prefer their leisure more than other workers. It may mean that the marginal product per hour of labour is particularly low as compared with other occupations and if it could be increased, more hours would be worked.

 TABLE 7.2 Hours worked by African cultivators

		Average hours per day	Average days per year	Average hours per year
Northern Nigeria, 1957[a]	Men			997
Calabar, Eastern Nigeria, 1956[b]	Men			1327
Malawi (Nyasaland), 1938[c]	Men			400–900
	Women			580–760
	Children			67–80
The Gambia, 1949[d]	Men and women		133	855
North Cameroons, 1958[e]	Men		106	
	Women		82	
Western Nigeria, 1967[f]	Men			1611
	Women			624
	Children			260
Ghana, 1955[g]	Men	4	174	696
Including fishing			190·5	762

Sources
[a]Baldwin, K. D. S. (1957) *The Niger Agricultural Project*, Oxford, Blackwell.
[b]Martin, A. (1956) *The oil palm economy of the Ibibio farmer*, Ibadan University Press.
[c]Platt, B. S. (1938) *Nutrition survey*, London, Colonial Office (mimeograph).
[d]Haswell, *Economics of agriculture in a savannah village*.
[e]Guillard, J. (1958) Essai de mesure, de l'activité du paysan africain: le Toupourri, *Agronomie Tropicale*, vol. 13, no. 4.
[f]Upton, *Agriculture in south-western Nigeria*.
[g]Lawson, R. M. (1968) The traditional utilization of labour in agriculture on the Lower Volta, Ghana, *Economic Bulletin of Ghana*, vol. 12, no. 1, p. 54.

Thus in the prosperous cocoa zone of Western Nigeria, farmers work the longest hours of any given in the table. The lowest figures occur in the poorer regions of Malawi and the Gambia. Alternatively the relatively few hours worked may be a reflection of the competing demands for the farmer's labour in secondary part-time occupations or non-farm work. Another major reason is the seasonal variation in farm labour requirements.

THE SEASONALITY OF FARM LABOUR REQUIREMENTS

Farm work, as has already been noted, does not occur evenly month by month throughout the year; busy periods or 'work peaks' alternate with slack periods or 'work troughs'. An example of a labour profile showing these peaks and troughs is given in Figure 7.5 which represents the monthly labour requirements for a mixed farm of 1·4 hectares (3·46 acres) in the Star/Kikuyu grass zone in Nyanza Province of Kenya.[7] The peak requirement for labour arises in the two planting periods; early planting in February–March and late planting in September. A smaller peak occurs in July at the time of sugar-cane harvest and uprooting of napier grass.

Work peaks occur because critical tasks such as planting, weeding and harvesting are closely related to the seasons and must be completed within a limited period of time. Delays generally cause loss of yield, so the man-hours needed to finish the task are compressed into a peak period. Other operations, particularly maintenance and repair work, allow greater flexibility of timing. Livestock work is usually spread fairly evenly through the year.

Unfortunately the farm labour supply can rarely be varied from day to day so as to match requirements exactly. The labour supply at busy periods may be increased by family members and hired men working extra long hours but this does not allow much flexibility. It is unlikely that temporary hired labour will be available just as and when required since landless labourers would prefer regular, secure employment and other farmers would only be available for temporary hire during the slack periods.

For the family work force and regular hired workers the supply of effort is relatively fixed throughout the year. It is therefore probable that the farmer will either have less labour than he wants on the farm at work peaks, or more than he wants at slack times, or something of both. Some seasonal unemployment or underemployment is almost inevitable in agriculture. This is illustrated in Figure 7.5 where the dotted horizontal lines show the total monthly labour supply for the farm family, assuming 2 adults each working 300 days annually. Even with the high supply

[7] Clayton, E. S. (1960) Labour use and farm planning in Kenya, *Experimental Agriculture*, vol. 28, no. 110, p. 83.

of 8 hours per day there is insufficient labour in February and September. Nevertheless there is underemployment in May, October, November and December.

In effect this means that the marginal product and hence the opportunity cost per hour of labour varies from season to season. At work peaks an extra hour of labour would yield a considerable increase in total product either because of more timely completion of the job or because a larger area may be covered. At slack periods, the marginal product of an extra hour of labour may be zero. The earlier discussion of the supply curve for effort, relating

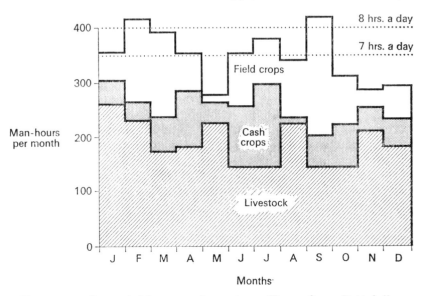

FIGURE 7.5 Seasonal labour requirements on Kenya farm. Dotted lines represent labour availability assuming 8 or 7 hours worked per day.

total hours worked per year to marginal product, is clearly an over-simplification. There is no single meaningful value for the marginal product or opportunity cost of labour which applies throughout the year. The seasonal variation in the opportunity cost of labour creates considerable complexity in farm planning and the management of labour.

To some extent the problem may be overcome by finding non-farm activities which will use labour during the slack farming periods. Members of the family may be able to organize their household duties and other tasks so that they are free to help on the farm at times when this help is most valuable. Secondary occupations such as house building, weaving, tailoring, pot and basket making, hunting and palm-wine tapping may be fitted in during the agricultural work troughs. Farmers may even migrate to find temporary employment during the season of agricultural underemployment. Similarly, farm improvements such as terracing, irrigation works, fencing and clearing new land may be carried out during the dry season when there is not much labour needed for farm crops.

However, as we have seen in considering what to produce on the farm, different enterprises may need labour at different times of the year. Where this is the case, the labour needs for a combination of enterprises will be more level through the year than the requirement for a single crop. For example, Figure 7.6 shows the monthly labour requirement per hectare of three crops grown in Kenya, namely early maize, late maize and coffee.[8] Clearly, early and late maize do not compete for labour because they use this resource at different times of the year. A combination of these two crops would spread the labour need over most of the year whereas either early maize or late maize on their own would leave labour unemployed for half the year. This implies that if a given regular labour force is needed to produce early maize, late maize could be introduced without any extra labour cost. The late crop would simply use labour in the slack period which would be otherwise unemployed or wasted. Late maize would then be a supplementary enterprise since it would supplement income without increasing the fixed labour cost.

More commonly enterprises compete for labour in some months but not in others. Thus from Figure 7.6 we see that coffee requires labour throughout the year. It therefore competes with early maize for labour during September, March, April, May and June. It does not compete with early maize, however, for labour in the months of July, August and October through to February.

Thus the allocation of labour between enterprises is a more complex problem than we have assumed so far, because marginal value products vary over the year. Labour cannot be allocated between enterprises simply by equating the overall marginal value products per man-hour. Ideally the marginal value products should be equated in every month of the year, or even each day of the year during which the enterprises compete for labour. Rarely, if ever, will the necessary data on the seasonal labour

---

[8] From Clayton, E. S. (1961) Economic and technical optima in peasant agriculture, *Journal of Agricultural Economics*, vol. 14, no. 3, p. 337.

response curves be available, so farmers must rely on past experience and personal judgement, leaving considerable scope for error.
The problem may be simplified by assuming, as in Figure 7.6,

Early (long rains) maize

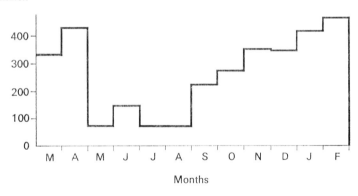

Late (short rains) maize

Man-hours
per hectare
per month

Coffee and mulch

Months

FIGURE 7.6 Seasonal labour requirements of early maize, late maize and
coffee per hectare. Early and late maize do not compete for resources; they are
supplementary. Maize and coffee only compete for labour in some months but
not others.

that the seasonal pattern of labour requirements per hectare of
crops and per head of livestock is *fixed*. Any reduction of labour at
peak periods would then necessitate a reduction in the area of
crops grown or the number of livestock kept. The opportunity
cost of introducing a new crop is then estimated in terms of the
areas of other crops or the number of livestock which would have

to be forgone. For example, in Figure 7.6 the work peak for early maize occurs in September when labour is also needed on the coffee crop. Hence the opportunity cost of increasing labour use on maize production can be calculated in terms of coffee forgone. From Table 7.3 it appears that the marginal return to an extra

TABLE 7.3 Returns to September labour

	I Man-hours per hectare in September	II Annual return per hectare (shillings)	III Annual return per man-hour in September (shillings)
Early maize	235	610	2·6
Coffee	225	4700	20·9

Note. Column III is obtained by dividing the annual return per hectare from column II by the September man-hours from column I.

man-hour of September labour in maize production is 2·6 shillings whereas the value of coffee forgone if 1 less man-hour is available is 20·9 shillings. Apparently, the opportunity cost of transferring each man-day of September labour from coffee production to early maize is much higher than the return. On this basis September labour should be used only for coffee production if profits are to be maximized.

However, this is not the whole problem, since work peaks occur on the coffee crop in February and April, so labour may be a critical limitation in these months too. A comparison of return per hour of labour on the two crops in these months might suggest a different allocation of labour. In this example at the prices given, coffee yields a higher return to labour in each month, but this need not always be the case. Possibly a combination of the two crops would be the best solution. The case is further complicated when other alternative enterprises are considered, each with a different return per hour of labour in each month.

This problem may be solved by linear programming (see Chapter 16) for very large numbers of alternative enterprises and large numbers of labour and other resource constraints, provided that the seasonal labour requirements are assumed to be the same for every unit of each enterprise. However, this is not a very realistic assumption, since there is some scope for varying labour inputs per hectare of crops and per head of livestock. The introduction of maize need not necessarily cause a reduction in the area of coffee. Instead the maize might be accommodated by reducing the hours worked per hectare on the existing area of coffee. These reductions would only occur in the peak months. However, without information on the response curve for coffee output upon labour input in specific months, the opportunity cost of such a policy cannot be estimated.

INCREASING LABOUR PRODUCTIVITY

Labour productivity, that is, the average value product per man-hour, may be increased either by producing more with the present labour supply or by saving labour. For the small family farmer it may be impossible to reduce the family labour force so labour saving would just cause unemployment. The only alternative is to increase total product. Where the population density is high and the amount of land available limits production, expansion can only be achieved by intensifying and increasing product per hectare, but where there is surplus land, labour productivity may be increased by extending the area under cultivation.[9]

Where there is sufficient land to extend cultivation the real constraint is not the total supply of labour but the amount available at the peak period. If this bottleneck can be eased or completely removed, some extension of the area under cultivation will be possible. Peak labour requirements may be reduced by modifying the farm system, changing the operations carried out, extending the period over which each operation is spread or by increasing the rate of working.

The scope for reducing work peaks by skilful choice of enterprises has already been discussed. In addition, having decided on a combination of crops and livestock, it may be possible to omit certain operations and thus reduce some work peaks. In particular the use of herbicides may eliminate the need for weeding and certain cultivations. The number of weeks over which each operation is spread also affects the productivity of the labour force at peak periods. An early start to clearing the land may enable a farmer to start planting his crops a week or a fortnight earlier than his neighbours. With the extra time available he should be able to plant a larger area of crops.

Finally, there are several ways of increasing rate of working on operations occurring at work peaks, including (1) the payment of

[9] See Sen. A. K. (1968) *Choice of techniques*, 3rd ed., Oxford, Blackwell.

incentives, (2) improving work methods and skills (method study) and (3) mechanization. These will be dealt with in more detail, though mechanization is deferred to the next chapter, but two general points are worth noting. Firstly, since the marginal product of labour is low at slack periods there would be no gain from increasing the rate of working at such times. Increased productivity can only be achieved by increasing rate of working on critical tasks occurring at work peaks, which must also include daily repetitive jobs such as fetching water, feeding livestock, milking cows and tapping rubber trees. Secondly, reducing the peak requirement for labour which sets a constraint upon output reveals subsidiary peaks, one of which in turn becomes the constraint upon output. Improvement in productivity is a continuing process in which the ideal is full employment throughout the season for the available workers.

INCENTIVES

Workers may be paid a regular hourly, daily or weekly time wage or they may be paid piecework rates in accordance with the amount of work done. Bush clearing is frequently carried out by piece-workers who are paid an agreed sum for clearing a particular area rather than so much per day. Piecework payments give a direct incentive to work at a fast rate since a man's earnings are directly related to the amount he can get done. However, piecework is more difficult to organize than labour hired at time-rates since each new task must be measured and acceptable rates of pay determined.

Crop sharing, which has already been mentioned as a form of land tenure, is quite common in Africa and may represent a form of incentive payment to labour, where that is the only resource provided by the share-cropper. This is the case already mentioned in Mid-Western state of Nigeria where rubber tappers usually receive half the rubber which they harvest as payment. They are in fact no more than hired labourers, but they receive a stronger incentive to maximize their productivity than if they were paid a fixed wage.

For some operations, quality of work may suffer from an increased rate of working. Land may not be properly cleared under piece-work rates, rubber trees may be carelessly tapped by share-croppers. Supervision of labour is often necessary, particularly in connection with the payment of incentives.

IMPROVEMENT OF WORKING METHODS

The worker's skill and the method used may have a significant influence on the rate of working and the quality of work done. Careful study of the way in which work is done and the time taken has led to significant improvements in many industries and, more recently, in certain agricultural operations.[10] For instance, 'work study' as it is usually called has led to improvement in methods of spraying fruit trees, lopping, weeding, and thinning root crops, pulling tobacco plants, stripping tobacco, picking tomatoes, apples, coffee beans and cotton, transplanting rice, milking cows and feeding pigs. A study of pig feeding recently carried out on a university farm in Nigeria resulted in a reduction in time and effort necessary for what appeared to be a simple task, performed several times a day.[11]

Normally the cost of employing a work study specialist can only be justified on large commercial farms, but even on a small farm it is worth while to look for ways of doing jobs more quickly and easily and of reducing wasted time. Furthermore there are some critical operations, such as hoeing, which create peaks in labour requirements on large numbers of small farms. The overall benefits from increasing the rate of working for this task on a large number of farms might be sufficient to justify a work study by a qualified researcher.

Work study consists of two main parts: work measurement and method study. The former is concerned with measuring rates of working and is useful in providing standards of labour requirements for specific operations and enterprises. The latter, method study, is aimed at finding improved methods of working. It consists of a systematic recording of the existing method followed by a critical examination and search for improvements. Method study is therefore only applied common sense but the systematic analysis of the work will give better and more reliable results than any haphazard procedure.

Two techniques of recording existing methods are the flow process chart and the string diagram. The flow process chart consists simply of a detailed listing of all the elements involved in a job. For instance the pig feeding operation on the Ibadan University

[10] For example see British Productivity Council (1962) *Work study in farming— twenty case studies*, London, B.P.C.

[11] Ogunfowora, O. (1969) The application of work study techniques in farm business management—a pilot study of the commercial pig fattening unit, University of Ibadan, *Bulletin of Rural Economics and Sociology*, Ibadan, vol. 4, no. 1, p. 52.

farm started with the following elements: (1) take trolley, (2) load one bag of ration, (3) wheel bag to mixing point, (4) open bag of ration, (5) empty feed into feed trolley, and so on.

Each element is then examined to ascertain firstly why it is done, secondly where it is done, thirdly when it is done, fourthly by whom and fifthly how it is done. Such critical examination may suggest the elimination or modification of certain elements in the job. Frequently the introduction of a simple piece of equipment such as a wheelbarrow or a larger handle on a hand tool may lead to considerable savings in time or energy.

The string diagram is particularly useful in illustrating wasteful movements and suggesting improvements in the layout of farm buildings or farm land. A plan of the work area is made on some convenient scale and all the movements made by the worker are then traced with thread, which is attached to pins inserted at each working point and each corner turned. The total length of the thread gives a measure, to scale, of the total distance moved, but the pattern made by the thread on the plan also emphasizes the pattern of movements. Frequently critical examination will suggest ways in which movement may be reduced, either by reorganizing the work, perhaps by cutting a new doorway or a new path, or by altering the whole farm layout.

Naturally, before a new method is introduced the extra costs must be weighed against the saving in labour or the increase in productivity. If the new method appears profitable there is then the problem of training workers to use it and of overcoming their resistance to change in their patterns of work.

SUGGESTIONS FOR FURTHER READING

BERG, E. J. (1961) Backward-sloping labour supply functions in dual economies— the Africa case, *Quarterly Journal of Economics*, vol. 75, no. 1, p. 468.

CLAYTON, E. S. (1960) Labour use and farm planning in Kenya, *Experimental Agriculture*, vol. 28, no. 110, p. 83.

CLAYTON, E. S. (1968) Opportunity costs and decision making in peasant agriculture, *Netherlands Journal of Agricultural Science*, vol. 16, no. 4, p. 243.

McFARQUHAR, A. M. M. (1959) *The importance of seasonal variations in the marginal value of labour on family farms*, Nigerian Institute of Social and Economic Research, Proceedings of 1959 Conference.

STURROCK, F. G. (1960) *Planning farm work*, London, H.M.S.O., Ministry of Agriculture, Fisheries and Food, Bulletin no. 172.

8 Capital

Capital is made up of things which, unlike natural resources, have been produced by human activity but which are not yet used up. For instance, a tractor is an item of capital; it has been produced but until it is finally scrapped it is not used up. Capital includes not only machines and tools but also buildings, roads, footpaths, drainage ditches, terraces, irrigation equipment, growing crops, livestock and stocks of food, seed, fertilizers and other materials. It should be clear that some capital is needed for any kind of productive activity. For instance the spears and food and water containers of pre-agricultural, food-gathering societies are items of capital. In fact, all a man's possessions and improvements to his land represent his capital. Each item of capital he owns is known as an asset.

We are concerned here with the assets used in the process of agricultural production but two other forms of capital should be noted in passing. One is social overhead capital, which includes communications, market places, public utilities, research stations and agricultural extension services, and is best considered here as a feature of the individual farmer's environment. The other is consumer capital, made up of durable consumer goods such as houses and furniture. It may be difficult to decide whether a particular asset is productive or consumer capital. For example, a bicycle may be used for pleasure or for transporting farm produce to market. A house, furniture and cooking utensils are necessities which must be available before a man is capable of productive work. The problems of definition are particularly acute in family farming where there are close links between farm and household.

Capital and money are often confused in everyday speech, but from an economic point of view it is important to distinguish between them, for money is only a convenient means of evaluating physical assets. In fact a commercial farmer will generally hold some of his capital in the form of money which can be used to buy stocks of seeds, fertilizers or other materials or to hire labour, but the money is not really productive until it is used in this way. For those capital assets which are produced on the farm it may be very difficult to estimate the cost of production although it is necessary to do so for rational decision making regarding the use of capital.

Capital is made up of so many different items that it can be misleading to treat it as a single resource. Land and labour vary in quality, as we have seen, but it is generally reasonable to assume that these resources can be transferred from one enterprise to another, say from cattle grazing to cotton production, without too much difficulty, although the marginal product per unit of land and labour may be very different in the two enterprises. With capital it is less straightforward. A cow may be worth say £30 which would be sufficient capital to grow 2 hectares of cotton; but a cow cannot be converted directly into the seeds, fertilizers and machinery needed to grow the cotton. Obviously buildings, terraces, irrigation works and other permanent structures cannot be readily converted into other forms of capital.

For purposes of planning the use of the existing stock of farm capital it is more realistic to separate different capital assets and measure them in physical terms. Machinery resources may be measured in available machine-hours just as labour resources are estimated in man-hours; buildings for storage or for livestock may be evaluated in terms of their capacity or floor space; numbers of livestock or hectares of irrigated land or of tree crops may serve as measures of the supply of these assets and stocks of food, seeds and other materials may be counted as physical quantities.

INVESTMENT AND DEPRECIATION

Investment refers to the production or acquisition of capital assets. It can take place in several apparently different ways, although in effect they all amount to the same thing, namely saving, which means forgoing current consumption. The first way in which a farmer invests is by actually saving some of his produce. For instance, cereals, legumes or yams which are stored either for seed or future consumption represent an addition to the stock of capital, and are therefore investments. Goats and cattle which are kept for milk or breeding or just to fatten into bigger animals also represent savings and investment.

Secondly, assets are created by the farmer's own physical

efforts. If he clears land, plants trees, builds a dam or a cattle kraal he is investing. Again he forgoes current consumption because he might have spent his time either in producing more food or in leisure, which in itself is a form of consumption. Investment by forgoing leisure is particularly suited to the peasant producer. Especially if his total output is little above the subsistence minimum he cannot afford to forgo consumption of produce, but even the poorest producer will almost certainly have some leisure time, particularly during the seasonal troughs in the pattern of labour requirements. The opportunity cost of this labour may be very small indeed.

The third method of investment is by purchase. This can, of course, only occur in an exchange economy where produce is sold or bartered, but this is true of most of Africa today. Again current consumption is forgone if the money or bartered goods would otherwise have been used for consumption purposes. Certain assets cannot be manufactured on the farm and must be purchased. This is particularly true of tools, machinery, stocks of improved seed and agricultural chemicals. It may occur to the reader at this point that capital assets are sometimes hired or purchased with the aid of a loan. We will return to these possibilities later but it is worth noting that hired or borrowed capital is still the outcome of saving by someone other than the user. Furthermore, although investment is ultimately dependent upon saving it does not follow that all savings are necessarily invested. They may be used for consumption, for festivities or just hoarded to be counted and gloated over from time to time.

Saving, no matter for what purpose, represents a cost to the user, namely the cost of waiting. Why then do individuals save? The answer is that the benefits that are obtained by waiting are greater than the value of the consumption forgone. In the case of productive investment, greater returns are obtained in the long run by using indirect or roundabout methods of production. Thus the farmer who spends time making a plough, training bullocks to draw it and destumping his land must work harder than his neighbours who use hoe cultivation, and he may produce less food than they while he is making this investment. In future years, however, he hopes that plough cultivation will add to his output more than enough to make up for his original efforts. This extra output, over and above the cost of the investment, is known as the 'return on capital'.

Each capital asset is eventually used up or destroyed; a stock of food is actually consumed, a cow or a cocoa tree dies, a tractor is broken up for scrap or a building falls down. Hence besides requiring a return on his capital the investor naturally hopes to recover the value of his original investment by the end of its productive life. In fact, if a farmer's operations are to continue on the same level he will need to replace his capital as it is used up. In other words a recurrent replacement cost will be associated with any new capital introduced into the farm system.

A farmer's capital assets may be classified according to the length of their productive lives into long-, medium- or short-term capital. Long-term capital has a life of many years and may be virtually permanent. It includes items like buildings, wells, dams and land improvements. Certain tree crops may come into this category. Capital with a medium-term life span of just a few years includes workstock such as bullocks, breeding and milking stock and many items of tools and equipment. Short-term capital is generally consumed within one year and includes stocks of food. seeds, agricultural chemicals and cash. The harvesting of annual crops usually occurs in one or two discreet periods of the year so not only do stocks of seed have to be provided some months before benefits are obtained but also family consumption must be met during the period between one harvest and the next. For these purposes capital in the form of stored seed and food (or the cash to buy them) is needed for at least part of the year. This short-term capital is also known as circulating or working capital to distinguish it from other assets which are not consumed within a single year and are therefore known as fixed capital.

In the case of medium- and long-term capital the replacement cost will not arise every year. Thus if the productive life of the bullock team and plough is estimated at five years, then maintaining the system will involve a large item for replacement every five years, which is not present in the intervening years. In many respects it is more convenient to think of the replacement cost as an imaginary annual series of cost items which would be equivalent in total burden to the five-yearly replacement cost actually involved. The imaginary annual cost of replacement is known as 'depreciation'. Alternatively we can think of depreciation as the annual loss in value of the asset, which is a more useful approach if the asset can be sold at any stage of its productive life. It should be noted that certain assets like livestock and trees actually gain in value or 'appreciate' over at least a part of their lives (see Figure 8.4, page 183).

One of the simplest ways of calculating an annual depreciation cost is to divide the original cost of the investment equally among the years of its productive life. Thus for a plough team costing £50 which lasts five years, the annual depreciation cost would be estimated at £50/5 = £10. This method is not entirely satisfactory because actual depreciation, meaning the annual decline in value, is not constant from year to year for most assets.

Although depreciation has been referred to as an imaginary cost, it may in fact be desirable to spread the necessary saving over the life of the asset. Otherwise the replacement cost, when it comes, may prove an embarrassment. From this point of view we should distinguish between gross investment and net investment. The former is the total investment made over a given period, say a year. Net investment, the actual increase in total capital, therefore equals gross investment minus depreciation over the same period. For example if a farmer keeps a flock of 10 goats but on average slaughters, sells or otherwise disposes of 3 each year, then he must breed or buy 3 goats each year as replacements to cover the depreciation of his flock. Only if he breeds or buys 4 or more goats in a particular year will he increase the size of his flock and thereby make a net investment.

For many items of fixed capital such as machinery, tree crops and livestock, there is the problem of how frequently to replace them. Should a tractor be replaced after five years or after ten years? Should tree crops be replanted after thirty years or after fifty years? This may be a very difficult decision in practice but the theoretical rule is that the asset should not be replaced until the extra return to be obtained from keeping it for one more year falls below the average return per year over its whole productive life.

The productive life of a capital asset may be prolonged by regular and careful maintenance so that the annual cost of depreciation is reduced. Repairs to buildings, drainage and irrigation works, servicing of machinery and pest control on trees and livestock will reduce the rate of depreciation. However, maintenance and repairs also involve costs so that a balance must be struck between maintenance costs and depreciation costs.

RETURN ON CAPITAL AND RISK DISCOUNTS

The annual return on capital is the total annual gain or benefit derived from using the capital, less all the extra costs incurred including depreciation, maintenance and repairs. It is generally expressed as a percentage of the total capital invested. As an example let us imagine a situation where it costs 10 bags of grain to make a plough and train the oxen to draw it; that is, the farmer has to forgo the consumption of 10 bags of grain or its equivalent. If the plough and ox team will need to be replaced at a similar cost in five years' time, then we may think of the annual depreciation as being $10/5 = 2$ bags of grain. Now, if there are no other extra costs and if the use of the plough team increases the farmer's annual production of grain by 4 bags the annual return on the capital is $4 - 2 = 2$ bags of grain, which represents 20 per cent of the original investment.

Clearly this is an oversimplified example. In the first place the total cost of an investment and the annual returns may be made up of many different items. Thus the introduction of draft oxen into an area hitherto cultivated by hand will involve major changes in the whole pattern of farming and possibly in the whole social structure of the community. In order to avoid frequent turns the land must be consolidated into fewer but larger fields; the ground must be cleared of termite mounds and tree roots and the oxen must be hand-fed while working. Indeed a change over to animal power may involve an overall increase in farm size. All these items must be evaluated in the same terms in order to estimate the total costs of the investment. Similarly the benefits will consist of increases in the output of several different crops grown on the farm and the oxen may yield output directly in the form of meat or young stock if they are used for breeding. Thus in order to compare and combine all these different items, it is usually necessary to evaluate capital and its return in money terms. Clearly when capital is invested in crop storage there is no gain in physical output; the return on capital depends upon increases in the product price.

The second simplification in the above example is that the salvage value of the plough and ox team has been ignored. Thus, if at the end of five years the oxen had a value as meat and could be exchanged for say 5 bags of grain, then the annual depreciation would only be $(10 - 5)/5 = 1$ bag of grain. The return on capital would then be 30 per cent.

Finally, we have made the simplifying assumption that the total amount of capital is constant at 10 bags of grain. In practice the costs of an investment are frequently spread over a period of time so that the total amount of capital needed gradually builds up to a

peak. Then as the returns start coming in, they may be set against
the capital cost so the total amount gradually falls. This variation
in the amount of capital involved in a productive activity may be
illustrated by considering the costs of establishing a tree crop such
as the oil palm. Estimates of the annual costs, excluding depreciation, and annual returns over the life of a hectare of oil palms
under Western Nigerian conditions are given in Table 8.1.

TABLE 8.1 Revenue and costs per hectare of oil palms

		£			
I	II	III	IV	V	VI
Year number	Yield (kg fruit)	Revenue	Costs	Margin	Cumulative cost (negative margin)
1	nil	nil	92·50	−92·50	−92·50
2	nil	nil	19·15	−19·15	−111·65
3	nil	nil	11·10	−11·10	−122·75
4	560	3·64	6·80	−3·16	**−125·91**
5	1680	10·92	8·75	+2·17	−123·74
6	2800	18·20	8·75	+9·45	−114·29
7	3360	21·84	10·18	+11·66	−102·63
8	4480	29·12	10·18	+18·94	−83·69
9	5040	32·76	10·18	+22·58	−61·11
10	5040	32·76	10·18	+22·58	−38·53
11	5040	32·76	10·18	+22·58	−15·95
12 to 35	5040	32·76	10·18	+22·58	positive balance

Note. The revenue per 1000 kg of fruit after deducting processing costs is £6·50,
so the revenue in the third column is obtained by multiplying the yield from the
second column by 6·50/1000. The margin in the fifth column is simply the
revenue (column III) minus the costs (column IV). The cumulative cost in
column VI is the total of all the annual margins up to and including that year.
After Agrawal, G. D. (1964) *Farm planning and management manual*, Ministry of
Agriculture and Natural Resources, Western Nigeria.

These estimates show a delay of three years between planting
and first harvest, then the yield rises to a maximum in the ninth
year and remains constant at this level for the rest of the productive
life of the trees. The trees are abandoned at the end of thirty-five
years and there is no salvage value. Costs of labour and materials
are incurred in establishing the trees in the first few years and in
maintaining them and harvesting the produce from then onwards.
As a result the cumulative cost of establishment rises to a peak of
approximately £126 in the fourth year, then gradually declines,

as shown in the last column of Table 8.1. The annual depreciation of the trees may be estimated by dividing the cost of establishment equally over the remaining thirty years of the crop life, to give an annual depreciation of £126/30 = £4·20. Thus from the fourth year onwards the capital value of the trees is assumed to fall by £4·20 per year down to zero at the end of the thirty-fifth year, as illustrated in Figure 8.4, page 183. The pattern of capital values over the life of a project, an asset or an enterprise is known as a 'capital profile'. The total capital requirement is determined by the peak requirement.

Requirements of working capital vary within each year according to a fairly regular cycle. Over the cropping season the costs of seed and of supporting the labour force must be met some weeks or months before the returns are obtained at harvest time. Thus the working capital requirement usually rises to a peak just before harvest. If the produce is stored after harvest the capital requirement continues to rise until the produce is finally sold or consumed. However, the peak capital requirements of different enterprises may occur at different times of the year. Where this is the case, the working capital requirement for a combination of enterprises will be more level through the year than the requirement for a single crop. Enterprises may be supplementary in the use of working capital in much the same way as they may be supplementary in the use of labour.

Because the amount of capital in use varies, there is some doubt as to which level should be used in calculating the percentage return. For the oil palm example given in Table 8.1, the annual margin of the mature crop from year (9) onwards is £22·58. By subtracting the annual depreciation of £4·20 we arrive at an annual return of £18·38. The percentage return on the peak capital requirement of £126 is therefore (£18·38/126) × 100 = 14·6 per cent. This rate of return will not be reached until the ninth year so it only applies to the last twenty-six years of the life of the crop.

However, £126 represents the maximum amount of capital in use. By the end of the life of the crop we assume that the value has depreciated to zero. Therefore, it might be argued that the average value of the trees over the whole productive life is the average of £126 and zero which is only £126/2 = £63. On this basis the *average* amount of capital is only £63 so the rate of return is (£18·38/63) × 100 = 29·2 per cent. Clearly this is a very different result from the rate of return on initial investment so it is important to decide which estimate should be used. For most purposes the rate of return on the initial investment is probably the most useful figure.

Because of these difficulties this approach of deducting a depreciation charge then relating the net annual return to the initial investment is not very satisfactory. Furthermore it makes no allowance for the cost of waiting. The delay between the initial investment and the mature return itself involves a cost which should be taken into account in estimating net return. Finally, this method can only be used where the annual margin is constant for a period of years after the investment matures. If it varied from year to year, the net return would vary as well as the amount of capital involved and it would be practically impossible to find a satisfactory single estimate of the rate of return. For these reasons time-discounting methods are increasingly used for appraising capital investments. However, the discussion of time-discounting is left until the appendix to this chapter; for the present we assume that return on capital can be calculated in the way just described.

Since there is a delay between the decision to invest in a particular activity and the receipt of the returns, capital investment always involves risk and uncertainty. Ordinarily the further ahead one is trying to predict, the greater the uncertainty will be. This means that for very long-term investments such as dams, irrigation works and other land improvements and buildings it is very difficult to predict the outcome in the more distant future. In fact the expected physical productive life might be indefinitely long but beyond a certain time horizon the predicted costs and returns are subject to such uncertainty that it is better to ignore them in estimating the return on capital. Thus for many long-term capital investments the return is calculated over a limited 'economic life' within which period the investment is expected to pay for itself. The choice of such a limited economic life is essentially arbitrary however.

A procedure for dealing with risk which is far more attractive than the limited economic life approach is the use of a risk-discount factor. For example, suppose the actual estimated rate of return is 10 per cent, then this might be reduced by a risk factor of, say, 1 per cent to a figure of 9 per cent for a mildly risky investment, or by a risk factor of 3 per cent to get a 'risk-discounted return' of 7 per cent for a more speculative project. It is this risk-discounted value of the prospective return which should enter into the calculations of the person concerned with a practical investment decision. For instance he may be faced with two alternative potential uses for his capital, say vegetable growing from which he would expect on average a net return of 15 per cent, and sorghum production yielding only 12 per cent. If however he thinks the former to be a far more risky proposition which should be risk-discounted by 5 per cent while the risk-discount for sorghum is only 1 per cent, then sorghum is the more attractive proposition (12 minus 1 per cent is greater than 15 minus 5 per cent).

The basic difficulty with discounting for risk is that there is no objective method of estimating what the risk factor should be and it must usually be estimated on the basis of past experience or some sort of judgement or intuition. In any event it must take into account the degree of the investor's dislike of risk which varies greatly between individuals.

When capital is defined in the broad sense used above, it is apparent that every agricultural activity requires a certain provision of capital since there is invariably a delay between the initial costs of starting production and the output of final product. Since the capital resources of a farmer are limited, they will restrict his range of choice as to the kind and level of productive activity he can pursue.

In theory the most profitable allocation occurs when the marginal value product is equal in each alternative activity, the marginal value product for capital being the extra risk-discounted return obtained by investing £1-worth (or some other small amount) more capital. However, many capital inputs are indivisible or 'chunky' so it is not possible to vary these by small or 'marginal' amounts. In practice it is not possible to adjust exactly to the point of equal marginal returns.

Nevertheless the principle of comparative advantage may still be applied; that is, to maximize profits each unit or chunk of capital should be used where it will earn the greatest risk-discounted return. This implies that the farmer should choose his enterprises or projects for investment in diminishing order of their risk-discounted returns. Once all the farmer's stock of capital is in use, the risk-discounted return on the last of these projects is the opportunity cost for any alternative investment. It should be noted again that it may be impossible to transfer fixed capital from an existing activity to a new one needing a different form of capital. Once the investment has been made certain costs are fixed, as discussed in Chapter 3.

THE COST OF SAVING AND THE SUPPLY OF CAPITAL

So far we have been considering the allocation of a fixed stock of capital but a farmer can, of course, increase his capital resources by saving. As we have seen, saving involves the cost of waiting; most individuals would prefer to receive a quantity of a particular commodity, say 10 bags of rice, *now* rather than the promise of the same amount a year from now. However, individuals vary in their time-preference; to be persuaded to wait a year one man may require the certainty of getting 11 bags of rice, while another would want 12. This means that the first man wants to be sure of getting 1 extra bag in 10, that is a 10 per cent return. The second man wants a 20 per cent return. Thus this personal time-preference sets a lower limit on the risk-discounted return a farmer will accept in deciding whether to invest. Naturally the higher a man values his present consumption in relation to future consumption, the higher this minimum acceptable rate of risk-discounted return will be.

It should not necessarily be inferred from this that if, because of improved technology or improved product prices, the risk-discounted return on capital rises, farmers will save more. The supply curve for savings may be backward sloping as is possible for labour. A rise in risk-discounted return increases the income of the investor and he may consequently prefer to increase his level of consumption and therefore to save less. Indeed the level of risk-discounted return probably has relatively little influence on the amount saved from any given level of income but much influence on the use of savings. The amount saved is likely to be affected much more by the level of farm incomes and the saving habits of the farming population. The saving habits of a person or a community are measured by the marginal propensity to save which is the proportion that is saved from each additional £1 (or other unit) of income.

It is usually found that rich men save more than poor men, not only in absolute amounts but also as a proportion of their total income. The very poor are unable to save at all. Instead they 'dis-save' or spend more than they earn, the difference being covered by going into debt or using up previously accumulated savings.[1] As incomes rise, so too does the marginal propensity to save. Thus if we compare African farmers with wealthier societies

[1] See Oluwasanmi, H. A. (1960) Agriculture in a developing economy, *Journal of Agricultural Economics*, vol. 14, no. 2, p. 234.

elsewhere we find that saving and capital investment are low because incomes are low, but it may be argued that incomes are low in turn because the amount of capital per person is low. This is known as 'the vicious circle of poverty' because it implies that poor people must remain poor unless capital is introduced from outside to break the circle. It also implies that poor farmers will have a greater preference for present consumption than will wealthier people. The poor farmers are more concerned with survival from day to day until the next harvest rather than with investment for the future. This means that they will only invest in activities which are expected to yield a relatively high rate of risk-discounted return. All we are saying really is that where capital is scarce in relation to labour and land, it will be costly in relation to these other resources.

So long as there is some net investment each year, no matter how little, the total stock of capital will increase continuously. This means that if there are diminishing returns to extra units of capital used with a fixed area of land and a fixed labour force, the rate of risk-discounted returns on the additional investments will fall over time. Eventually the rate of return will fall to a minimum acceptable level and no further investment will be justified. Thus a point of stagnation will be reached where farmers are capable of acquiring more capital by saving but where there are no further opportunities for productive investment.[2]

In practice the situation is complicated by the process of change. As shown in the previous chapter some net investment is necessary to keep pace with population growth even if there are no innovations; but where agricultural change is taking place, new investment opportunities are constantly being introduced. Such technical innovations generally yield higher risk-discounted returns per unit of capital than do extra investments in traditional productive activities. Thus technical innovations provide new opportunities for productive investment, to which farmers may respond without outside assistance.

A good example is the rapid expansion of cocoa production in West Africa during the last half-century. The massive investment in establishing the trees was provided by the small farmers themselves as they became aware of the possibility of growing this new crop for export. Similarly in Kenya more recently the establishment of tree crops was largely financed by the farmers

---

[2] T. W. Schultz has explored this idea of the stagnation of traditional agriculture in his book *Transforming traditional agriculture*, Yale University Press, 1964.

themselves.[3] In fact, among cash-cropping farmers the rate of saving and investment is often high by any standards. Studies among cocoa farmers of Nigeria have suggested that they may save as much as one-third of their incomes.[4] Thus the introduction of new techniques and products can enable small farmers to break out of the vicious circle of poverty.

NEW FORMS OF CAPITAL

Most technical innovations are embodied in new forms of capital. They may be broadly classified into product innovations, which are new crops or livestock introduced to an area, and process innovations, which are new methods of producing existing products. Thus the introduction of fertilizers and other agricultural chemicals, irrigation and other land improvements, mechanization and improved tools, equipment and buildings are all basically process innovations. Clearly product and process innovations may be introduced at the same time and it is difficult to classify some items such as new varieties of existing crops.

Process innovations are further classified according to whether they increase the productivity of all resources equally, in which case they are called neutral and output-raising, or save one particular resource rather than another. For instance mechanization is usually a labour-saving innovation, whereas irrigation, drainage, conservation measures and fertilizers are land-saving. It is even possible for new forms of capital to be capital-saving innovations. Thus a new chemical spray which controls disease on cocoa trees and thereby extends their productive life is capital-saving; so too is a modern grain store which reduces losses from stored grain, or a rust-preventive that makes machines last longer. The distinction between labour-saving, capital-saving and neutral innovations is illustrated by the set of hypothetical isoquants shown in Figure 8.1. If the relative prices of labour and capital remain unchanged, the introduction of a neutral innovation saves both labour and capital, needed to produce a given level of output, in the same proportion.

These distinctions are very important in relation to the transfer of techniques from other parts of the world where relative factor costs may be different. Thus, in comparison with the agriculture

[3] See de Wilde, J. C. and others (1966) *Experiences with agricultural development in tropical Africa*: vol. 1, *The synthesis*, Baltimore, Johns Hopkins Press.
[4] See Upton, *Agriculture in south-western Nigeria*.

of Western Europe and America, the price of labour in African farming is low and the price of capital is high. This means that labour-saving farm machinery will produce a lower rate of return on most African farms than it will in Europe or America. In Africa, such machinery may increase the unemployment problem. Generally speaking labour and land are relatively plentiful and therefore cheap in relation to capital. This suggests that neutral and capital-saving innovations may yield the highest return on capital.

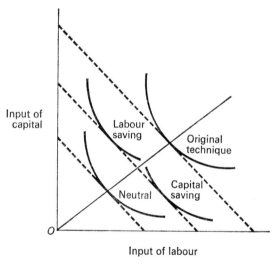

FIGURE 8.1 Labour saving, capital saving and neutral innovations. The new techniques reduce the amounts of labour and/or capital needed to produce a given level of output. Hence the isoquants are shifted down towards the origin. If the ratio of labour to capital stays the same it is a *neutral* innovation.

However, in practice there are very few innovations which save only one resource. Land-saving and labour-saving innovations also increase total product output. Tractor mechanization, for instance, not only saves labour but also increases yields, as suggested by the results from the Mwea-Tebere irrigation scheme given in Table 8.2. Yield increases may result from more effective cultivation than is possible by hand, or from earlier completion of major tasks of planting and hoeing. Furthermore tractors enable the cultivation of heavy soils which cannot be handled with hoe or oxen. Apart from the saving in costs or increases of output, innovations may yield benefits in terms of risk reduction. Irrigation reduces the risks of crop failure through drought, chemical sprays reduce the risks of product losses from pest or disease attacks, mechanization reduces risks of failure to complete critical peak operations in time. This suggests a need for a so-called 'intermediate technology', made up of techniques and equipment which will increase output and reduce risk but which are less laboursaving and less costly than many of the methods used in countries where relative labour costs are higher.[5]

TABLE 8.2 Mechanization and crop yields: yield in tons per acre of cleaned winnowed paddy at 14 per cent moisture

		Yield	Increase per cent
1959–60	Ox-cultivated crop	1·74	
1960–1	Partial tractorization	2·16	24
1961–2	Fully tractorized	2·29	32
1962–3	Fully tractorized	2·40	38

Source: Giglioli, E. G. (1965) Mechanical cultivation of rice in the Mwea Irrigation Settlement, *East African Agricultural and Forestry Journal,* January 1965.

Nevertheless, it is dangerous to generalize regarding the likely returns from different types of investment. Ideally, the agricultural planner should estimate the return on capital to be expected from each alternative investment opportunity in the particular environment with which he is concerned. Some estimates of the percentage return on capital from different agricultural projects in Kenya are given in Table 8.3. It should be noted that these rates of return are not risk-discounted and the estimates are based on fragmentary evidence, so they must be interpreted with care. However they do serve as an illustration.

CREDIT

The principal external sources of capital are gifts and inheritance on the one hand, and loans or credit on the other. Gifts and shortterm loans are of considerable importance within the extended family, providing a basis for transferring income from one person to another in rotating situations of crisis and need. Yet family sources are often inadequate and farmers have to borrow from

[5] See Intermediate Technology Development Group (1967) *Tools for progress,* I.T.D.G., 9 King Street, Covent Garden, London W.C.2. Also *Intermediate Technology Development Group Quarterly Bulletin.*

other sources either following upon a failure of crops, a fall in prices, illness and natural calamities or to meet expenditure out of proportion to income on the performance of ceremonies dictated by custom and public opinion. Such loans are used for consumption rather than for productive investment so they do not provide any increase in income out of which to repay the lender. As a result of such unproductive borrowing some farmers become deeply indebted and dependent on moneylenders.

TABLE 8.3 Economic returns from alternative development projects in Kenya

Project	Internal rate of return per cent
Smallholder tea development	50
Agricultural administration and extension	
Nyeri district	40
Elgeyo-Marakwet district	35
Central Nyanza district	12
All three districts	29
Irrigation development	
Mwea-Tebere scheme	22
Mwea-Tebere and Perkerra schemes together	18
Low density resettlement	
90 per cent fulfilment of targets	9
70 per cent fulfilment of targets	6
New settlement on rain-fed land	4

Note. For details of the calculation of internal rate of return see the appendix to this chapter.

After Ruthenberg, H. (1966) *Agricultural production development policy in Kenya, 1952–1965,* Afrika-Studien no. 10, Berlin, IFO Institut.

Other loans are used for productive investment and provide a return great enough to pay all expenses of servicing the loan and still leave a surplus. The use of such productive credit should be viewed as a normal part of efficient farm operation. Indeed some new techniques require such large increases in the capital invested that even with high rates of saving, small farmers are unlikely to be able to afford them without the use of credit. Some estimates of the capital investment on traditional farms and in various improvements, given in Table 8.4, illustrate how large the increases may be.

TABLE 8.4 Capital resources on African farms

	Type of capital	Estimated value ($£$)
Yoruba cocoa farmers, 1952[a]	Tools and equipment	6
	Cocoa trees	550–600
Ibibio oil palm producers, 1955[b]	Tools	2
Zaria Hausa, 1950[c]	Tools	0·18–0·85
Ibo rubber farmers, 1964[d]	Tools and equipment	5
	Livestock	21
	Tree crops	424
Mixed farming settler, N. Nigeria[e]	Land and improvement	900
	Roads and services	100
	Buildings	220
	Livestock	75
	Implements	65
Mweiga settlement, Kenya[f]		357
Ainabkoi settlement, Kenya[f]		662
Ilora farm settlement, W. Nigeria[g]		5000

Sources
[a]Galletti, Baldwin and Dina, *Nigerian cocoa farmers*.
[b]Martin, A. (1956) *The oil palm economy of the Ibibio farmer*, Ibadan University Press.
[c]Smith, M. G. (1958) *The economy of Hausa communities in Zaria*, London, H.M.S.O., Colonial Research Study no. 16.
[d]Upton, *Agriculture in south-western Nigeria*.
[e]Alkali, M. M. (1967) Mixed farming in Northern Nigeria, in *Problems and approaches in agricultural development*, part of Proceedings of Joint D.S.E.–E.C.A.–F.A.O. Seminar.
[f]de Wilde, J. C. and others (1967) *Experiences with agricultural development in tropical Africa*: vol. 2, *Case studies*, Baltimore, Johns Hopkins Press.
[g]F.A.O. (1966) *Agricultural development in Nigeria 1965–1980*, Rome.

INTEREST AND SECURITY FOR LOANS

An individual who lends capital to someone else usually expects some payment in return. This payment, the cost of hiring capital, is generally expressed as a percentage of the total loan and is known as 'interest'. To the lender it is repayment for the cost of waiting. Even where the lender is a government organization, a bank or a big corporation any capital which is made available is the outcome of someone's saving. Hence all capital has a cost.

The charging of interest, or usury as it used to be called, is condemned in the Bible and the Koran and in Marxist ideology, but it does serve two useful purposes. Firstly it is an incentive to individuals and organizations to lend capital. Secondly it induces the most productive allocation of available capital, provided that interest rates are allowed to find their competitive level, in much the same way as rents ration land. Thus the enterprises yielding the highest risk-discounted returns will be able to pay the highest interest rates and hence will have first call on available capital.

When capital is borrowed, interest is the cost of the capital which must be compared with the risk-discounted return in deciding whether to invest. In situations where farmers can borrow or lend money at interest, the market rate of interest represents the opportunity cost for the farmers' own capital. If the rate of interest exceeds the risk-discounted return a man can earn on his own farm he will benefit by lending to others.

When capital is lent there is always a risk that the loan will not be repaid. Indeed if the project fails or if the loan is used unproductively the borrower may be unable to repay his debt. Hence the lender generally wants some reassurance against the complete loss of his capital. On the one hand this reassurance may be provided by a personal knowledge of, and trust in, the borrower. Such reassurance may well be adequate between members of the same family, or village community. On the other hand, where relationships are less personal the lender will require some con-tractual assurance of repayment.

Contractual assurances may take various forms. Thus merchants and statutory buying agencies often provide short-term credit services. The possibility of deducting repayment of loans from the price for the indebted farmer's produce may be adequate assur-ance. For longer-term loans the lender requires some 'collateral security' for his loan. This collateral security is some asset which, in the event of failure to repay the loan, becomes the property of the lender. For instance, a tractor or some other machine bought by hire-purchase is itself the security for the loan. Land, trees and buildings offered as security for loans are said to be mortgaged. In contrast assets which are 'pledged' are actually used by the creditor (the person providing the credit) until the debt is repaid. Since most African farmers own very few movable assets and do not own their land in any legal sense they have very little to offer as security for long-term loans.

BORROWING AND RISK

Not only the lender but also the borrower faces the risk of losing his capital. Indeed the greater the proportion of his total assets a farmer borrows, the greater is the chance of his losing his own capital, as shown by the hypothetical figures in Table 8.5. The value of the capital actually owned by the farmer is known as his 'net worth' or 'equity'. This expressed as a percentage of his total capital both owned and borrowed is known as his equity ratio. As a farmer borrows more money, he reduces his equity ratio and increases the variation in his residual profit margin. This is because

TABLE 8.5 Borrowing and risk: the effect of increased borrowing on the farmer's profit margin

	Equity ratio		
	100%	50%	25%
	£	£	£
Own capital used	200	200	200
Borrowed capital used	0	200	600
Total capital used	200	400	800
Good years (return on capital 20%)			
Total return	40	80	160
Interest on borrowed capital at 10%	0	20	60
Residual profit margin	40	60	100
Expressed as return on own capital	20%	30%	50%
Bad years (return on capital 5%)			
Total return	10	20	40
Interest on borrowed capital at 10%	0	20	60
Residual profit margin	10	0	−20
Expressed as return on own capital	5%	0%	−10%

the annual interest charges on the borrowed money do not vary with the year to year fluctuations in the farmer's income. In good years when the return on capital is greater than the interest rate, the farm profit is increased by borrowing. But in bad years when the return on capital falls below the interest rate, profit is reduced by borrowing. Indeed in such years a farmer with a low equity ratio may find he owes more in interest payments than he has earned in income, as shown in the table. In order to pay the interest he must either draw on his own capital or get deeper in debt, either of which will further reduce his equity ratio.

These risks associated with borrowing may discourage farmers from making major changes in their pattern of farming which will involve them in considerable debt.

PRIVATE LOANS AND MONEYLENDERS

The major proportion of all credit used in rural Africa comes from local sources, the bulk from relatives and friends. In most such cases little or no interest is charged, loans are usually short-term and no security is required. Local moneylenders, many of whom are traders, may make small cash advances on the understanding that the borrower will sell his crops to the lender. Besides giving the lender a chance to recover his loan after harvest this also assures him in advance of his supply of produce.

It is widely believed that the rates of interest charged by local moneylenders are excessive. Indeed rates of over 100 per cent per year are sometimes paid, which may be contrasted with rates of around 10 per cent normally charged by banks and official credit organizations. However, it should be remembered first that in rural areas capital is scarce, hence the opportunity cost is high, second that the administrative costs of handling many small loans is high, and third that the moneylender has very little real security and the risk of default on repayment is high. On the other hand, if there is a single wealthy trader in the district who is the sole source of credit he can resort to extortion and charge excessively high rates of interest. Thus there may be a case for the government to provide rural credit on grounds of social justice as an alternative source. Even then it is better not to eliminate the private source of credit. With two or more sources of credit, each one bids for the farmer's patronage by making its interest rates and its method of making loans as reasonable as good financial practice can allow. Consequently, farmers are less likely to feel that they are being exploited by the lender with whom they choose to deal, whether he is a private individual or a government agent.

It should be added, though, that for short-term loans local moneylenders or merchants may be very attractive to farmers despite the high interest rates, the main attractions being the lack of need to provide collateral security and the ease of getting such loans, which may be made on the spot without any formal contract.

In high income countries and among industrialists in Africa the main sources of credit are the commercial banks (which have of course been nationalized in some countries). However, because of the administrative problems of lending to farmers and the absence of collateral security, these banks generally do not provide much agricultural credit.

GOVERNMENT AND COOPERATIVE CREDIT

Local sources of credit, either friends and neighbours or traders and moneylenders, may be adequate sources of credit for short periods of up to a year. The problems of providing medium- and long-term credit for development are more difficult. Merchants are rarely willing to lend for longer than one cropping season against the assurance of the harvested crop, and as we have seen, small farmers have very few assets to offer as security for loans.

The total size of a farm business may be increased, where this is thought to be beneficial, by the pooling of capital resources. This is one reason for the formation of producers' cooperatives, whereby a group of producers all contribute to a common pool of capital. The possibilities range from the sharing of a simple machine such as a groundnut huller, to a fully integrated communal farm on which the members share the work, the costs, the risks and the proceeds. Membership can range from as few as five up to a very large number, although as the organization gets bigger, management becomes more complex and hence more remote from the individual members. The original cooperative spirit may be lost.

It is usually thought desirable that each member should contribute roughly the same amount of capital so that no individual can come to dominate the cooperative. The common pool of capital is organized and made productive by the cooperative manager who may be elected from among the membership of a small cooperative, or who is more likely to be a paid employee of a larger one. In some cases there is a management committee. The money surplus is usually distributed in proportion to the amount the cooperative is used. The members should have the right to displace an inefficient or corrupt officer. However, they usually need some supervision by government officers, since the members have inadequate experience of running the affairs of an organization and looking after its money. Indeed in many instances the form of administration of cooperatives is rigidly controlled by government regulations and sometimes these organizations are actually administered by government agencies.

Producers' cooperatives have been set up or encouraged as part of government policy in various parts of Africa but on the whole cooperation on the production side has been less successful than in the field of marketing. This may be because the economies of scale are limited in African agriculture or that the spirit of cooperation is not strong enough to overcome the individualism and independence of the African farmer, at least in his farming activity.[6]

Cooperative credit agencies are playing an increasing role in Africa. Basically the administrative principles are the same as for a producers' cooperative but the objective here is to pool capital savings for the purpose of lending to individual members for productive use. (Cooperatives being non-profit-making organizations generally keep interest rates down to the minimum necessary to pay all costs.) Clearly without outside assistance they do not raise the total capital resources of all the members, *but* many governments favour providing funds to farmers through cooperative agencies. The cooperatives have certain advantages in personal knowledge of members applying for loans and may take joint responsibility for government funds. Some cooperative credit agencies have grown very big, even becoming national cooperative credit banks, but then the close personal association is lost. Large cooperative banks tend to be administered by bureaucrats and the procedure for getting loans may become slow and tedious.

This latter problem may arise with government-sponsored agricultural development banks or other credit agencies. In particular, time is needed for careful screening of each individual applicant for loans especially where there is inadequate collateral security. In fact it has frequently proved beneficial for the credit agency to work closely with the agricultural extension service, making use of their local knowledge of individual farmers. Frequently these government credit agencies keep the rate of interest down below the cost of the capital including its administration and risk. In other words agricultural credit is subsidized, with the objective of encouraging farmers to borrow and invest in new products and practices. Such schemes have met with varying success. In some cases loans have been used for unproductive purposes and farmers have defaulted on their repayments.[7]

On irrigation schemes and other agricultural settlements, capital is provided to the settlers in the form of improvements to the land, irrigation works, roads, buildings and so on. The capital is

[6] See Whetham, E. H. (1968) *Cooperation, land reform and land settlement*, London, the Plunkett Foundation for Cooperative Studies.

[7] See Hunt, D. I. (1966) *Some aspects of agricultural credit in Uganda*, Makerere University College, Economic Research Development Paper no. 105 (mimeograph); *The operation of the progressive farmers' loan schemes in the Lango district*, Makerere University College, Rural Development Research Paper no. 30 (mimeograph); also Harrison, A. (1967) *Agricultural credit in Botswana*, University of Reading, Department of Agricultural Economics, Development Study no. 4.

provided on loan initially although in many cases the settlers are expected to repay the loan with interest. These repayments place a heavy burden on the settlers in the early stages of a scheme. Alternatively, if the settlers remain as tenants, the contract rent covers the interest and depreciation costs of the capital invested.

On irrigation schemes, certain costs of providing irrigation water vary with the quantity of water used. From the point of view of encouraging the most profitable use of available water the scheme authority or owner should charge a water rate related to the variable cost of providing water, apart from the fixed costs incorporated in the rents or repayments of the initial investment.

SUPERVISED CREDIT AND HIRING

Supervised credit is productive credit which is offered in conjunction with technical advice and assistance. The credit agent, who must be a trained agricultural extension worker, first helps the farmer to make a production plan for his farm for the coming year. It includes an estimate of the amount of credit needed to finance the plan and the probable value of the increased product. Credit is then provided either in cash or in the form of the specific supplies and equipment needed. The credit agent visits the farmer from time to time, giving technical advice and checking that the farmer is following the plan drawn up. In some cases new inputs such as new seeds, fertilizers or machinery services are offered.

Thus the credit and the technical assistance are complementary to each other. The credit ensures that the farmer can finance the new techniques and these in turn ensure sufficient increase in income to repay the loans with interest. The close supervision ensures that credit is used productively.

Such schemes are, of course, costly. Few farmers can be supervised by each credit agent, who must be highly skilled and well trained in the technical and management problems of farming. However, the combined package of innovations and credit may be more productive than individual services offered alone. Major supervised credit schemes are in operation in Brazil, Mexico and India.[8]

Hiring should also be mentioned here as a form of credit particularly applicable to machinery and possibly livestock. Where a machinery hire service is intended to be self-supporting the charges

[8] See F.A.O. (1964) *New approach to agricultural credit*, Agricultural Development Paper no. 77, Rome.

must cover the costs not only of repairs, maintenance, depreciation and operator wages but also interest on the capital invested. Livestock hiring schemes exist in some parts of Africa often with interest payments being paid in kind. For instance in south-eastern Ghana, Fulani herdsmen tend cattle owned by members of the local Ewe tribe. The calves which are returned to the owners may be viewed as a hire charge.

SUGGESTIONS FOR FURTHER READING

BELSHAW, H. (1959) *Agricultural credit in economically underdeveloped countries*, Rome, F.A.O.
BROSSARD, D. B. and GRETTON, R. H. (1962) *Agricultural credit and cooperatives in rural development*, Rome, F.A.O.
HARRISON, A. (1956) The 'capital profile' as an aid to decision making in farm management, *Journal of Agricultural Economics*, vol. 12, no. 1, p. 64.
HAWKINS, C. J. and PEARCE, D. W. (1971) *Capital investment appraisal*, London, Macmillan.
HILL, P. (1970) *Studies in rural capitalism in West Africa*, African Studies Series no. 2, Cambridge University Press.
McFARQUHAR, A. M. M. (1965) *Mechanization in agriculture in underdeveloped economies*, Paper to the British Association for the Advancement of Science.
MERRETT, A. J. and SYKES, A. (1963) *The finance and analysis of capital projects*, London, Longmans.
SCHULTZ, T. W. (1964) *Transforming traditional agriculture*, Yale University Press.

Mathematical appendix to Chapter 8

COMPOUND INTEREST, THE COST OF WAITING

The delay between the input of capital and the receipt of its product complicates the estimation of return on capital. As we have seen the value of a capital asset varies over its productive life so problems arise in defining the amount of capital in use. Furthermore the net return may vary from year to year over its life. Finally the delay involves the cost of waiting which should be taken into account. These problems may be overcome by the techniques of compounding and discounting.

Compounding means adding interest charges on the cumulative debt at the end of each year. This may be illustrated with an imaginary example of a timber plantation established for an initial cost of £30. Twenty years later the trees are sold for £300. Now, if the farmer borrows the initial £30 at 10 per cent interest, we can estimate the total debt year by year over the life of the plantation. By the end of the first year the total debt will amount to £30 plus £3 interest. The farmer has not yet received any returns so he must borrow this too. In the second year he will again have to borrow the interest but this time it will be 10 per cent of £33 which equals £3·30. Hence the total debt at the end of the second year will be £36·30. Similarly we could go on calculating the debt year by year up until the end of the twentieth year. This is rather a tedious procedure but fortunately we can derive a formula for estimating the cumulative debt owed at the end of any number of years and at any rate of interest.

Let P = original loan or principal,
r = interest rate expressed as a decimal e.g. 10 per cent
$= 0·1$.

Then the debt after 1 year V_1 is the principal plus interest:

$$V_1 = P + r \cdot P = (1 + r)P.$$

After 2 years the debt is:

$$V_2 = (1 + r)P + r(1 + r)P = (1 + r)^2 P.$$

After 3 years the debt is:

$$V_3 = (1 + r)^2 P + r(1 + r)^2 P = (1 + r)^3 P.$$

In fact the debt at the end of any number of years may be calculated by the general formula:

$$V_n = (1 + r)^n P = P(1 + r)^n \qquad (1)$$

where n = the number of years which have elapsed since the original loan.

To complete the example of the timber plantation we estimate the cumulative capital cost at the end of 20 years by compounding the original £30 loan:

$$V_n = £30 \, (1 + 0·1)^{20} = £30 \, (1·1)^{20} = £201·80.$$

Thus at a 10 per cent rate of interest there is a positive profit since the total cost will be £201·80 and the total return £300. The profit to be made in 20 years' time is £98·20.

The profitability of the investment is highly dependent on the interest rate. At 15 per cent the above project would produce a large loss, since

$$V_n = £30 \, (1·15)^{20} = £491.$$

A return of £300 now results in a loss of £191.

In order to reduce the arithmetic involved in calculations such as these, Table I has been included at the end of the book, showing the future value of £1 at different interest rates and for different periods of time. We need only multiply the future value indicated, by the original sum to complete the compounding exercise. Thus we see from Table I that the future value of £1 compounded at 10 per cent over 20 years is £6·73, so the future value of £30 is 30 × £6·73 = £201·90.

In practice the costs may be spread over a period of several years, in which case the cost for each year must be compounded separately. Thus the costs in year (1) must be compounded over 20 years whereas those in year (2) will only accumulate compound interest over 19 years, and so on. Furthermore, if revenue is obtained some years before the end of the productive life of the investment, it will reduce the total debt which is outstanding and therefore the interest payments. In effect the revenue earns interest by saving interest costs. For this reason revenues which are received before the end of the project life should also be compounded. Thus we compound all 'cash flows' whether they are costs or revenues.

The technique applies equally when the farmer's own capital is used, provided that we know his personal time preference or the opportunity cost of his capital. Thus if he values 10 bags of grain now as equivalent to 12 bags of grain 1 year from now, he wants a 20 per cent return. By the same token if his time preference remains unchanged he will value 12 bags 1 year from now as equivalent to $12(1·2) = 14·4$ bags 2 years from now. This is the same as charging 20 per cent interest per year.

ANNUITY

An important special case arises if a series of equal cash flows are expected every year over a period of years. For instance a plough team or a milk cow might yield approximately the same benefits each year over its productive life. Such a series of equal cash flows is known as an annuity.

If we let A = the annual cash flow then the future or terminal value V_{an} is calculated as follows:

$$V_{an} = A\,(1 + r)^{n-1} + A(1 + r)^{n-2} + \ldots + A(1 + r) + A$$
$$= A[(1 + r)^{n-1} + (1 + r)^{n-2} + \ldots\ldots + (1 + r) + 1].$$

Multiplying both sides by $[(1 + r) - 1]$ gives

$$V_{an} [(1 + r) - 1] = A[(1 + r)^n + (1 + r)^{n-1} \\ + \ldots + (1 + r)^2 + (1 + r)] \\ - A[(1 + r)^{n-1} + \ldots + (1 + r)^2 + (1 + r) + 1].$$

Therefore:

$$V_{an} \cdot r = A [(1 + r)^n - 1]$$

so:

$$V_{an} = \frac{A[(1 + r)^n - 1]}{r}. \tag{2}$$

For example, if a plough team is expected to yield an extra return of 2 bags of grain worth £2 each year for the next 5 years, the value of this extra produce at the end of the 5-year period, assuming a 20 per cent or 0·20 rate of interest, is given by:

$$V_{an} = \frac{£2 \left[(1·20)^5 - 1\right]}{0·20} = £14·88.$$

Again the arithmetic is reduced by using tabulated future values of a £1 annuity at different rates of interest and for different periods of time (see Table II at the end of the book). This table can also be used to calculate the sum that must be set aside each year in order to accumulate a given total amount at a specific date in the future. For instance if we want to replace £43 at the end of 13 years and the rate of interest is 8 per cent, we find that an annuity of £1 for 13 years at 8 per cent yields a terminal value of £21·50. Thus to replace £43 then £43/21·50 = £2 must be invested each year at 8 per cent interest for 13 years. In this way, an annual depreciation charge may be estimated as the annual sum that must be saved and invested. It is then called a 'sinking fund'.

DISCOUNTING

If a farmer will only invest 10 bags of grain now if he expects to gain 12 bags of grain 1 year from now, this means that his 'present value' of 12 bags of grain expected 1 year hence is 10 bags of grain. Likewise if the interest rate on borrowed money is 10 per cent and at this rate a cost of £30 accumulates to £201·80 in 20 years, it means that the 'present value' of £201·80 at 10 per cent interest is £30 (see Figure 8·2). This process of estimating the present value of future cash flows is known as discounting, and is simply the converse of compounding.

Thus if:

$$V_n = P(1 + r)^n$$

then if we divide both sides by $(1 + r)^n$ we obtain

$$P = \frac{V_n}{(1 + r)^n}. \tag{3}$$

We are now discounting for time not for risk. However, we may add a risk-discount to the rate of interest (r) to allow for risk. Table III at the end of the book gives present values of £1 dis-

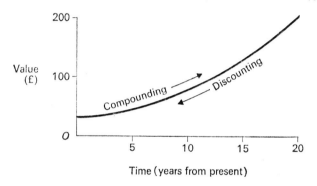

FIGURE 8.2 Discounting, the converse of compounding. If £30 compounded over 20 years at 10 per cent will be worth £201·80, then £201·80 discounted over 20 years at 10 per cent is worth £30 now.

counted at different rates of interest over different numbers of years. Thus the cash flow for a particular year is simply multiplied by the value in the table for the relevant number of years and rate of interest to give the present value.

In practice, discounting is generally more useful than compounding because the farmer is concerned with present values rather than future values in making current decisions. Thus, for the example of the timber plantation we have used to demonstrate compounding, it is not very helpful to know that the profit in 20 years' time is expected to be £98·20. However, this can be discounted at 10 per cent to give a present value as follows:

$$P = \frac{V_n}{(1 + r)^n} = \frac{£98·20}{(1·1)^{20}} = £14·60$$

or from Table III as £98·20 × 0·1486 = £14·60. Since this is the difference between cost and revenue it is known as the net present value (NPV).

To arrive at this estimated net present value we have compounded costs, then discounted the final profit at the same rate of interest. It would have been more straightforward to discount the revenue to start with, rather than to compound the initial cost. A revenue of £300 discounted at a rate of 10 per cent over 20 years gives a present value of £44·60. Then, by subtracting the initial cost of £30 we arrive at the net present value of £14·60.

In real life, costs and revenues are more likely to be spread over several years to give a series of cash flows.

Let V_1 = cash flow at end of 1 year,
V_2 = cash flow at end of second year, and so on up to
V_n = cash flow at end of final year of the project life.

Then:

$$P = \frac{V_1}{(1+r)} + \frac{V_2}{(1+r)^2} + \frac{V_3}{(1+r)^3} + \cdots + \frac{V_n}{(1+r)^n}.$$

DISCOUNTING AN ANNUITY

When the cash flows represent an annuity we have:

$$V_1 = V_2 = V_3 = \ldots = V_n = A$$

so that

$$P_{an} = A\left[\frac{1}{(1+r)} + \frac{1}{(1+r)^2} + \frac{1}{(1+r)^3} + \cdots + \frac{1}{(1+r)^n}\right].$$

This can be simplified as follows: by using the formula for calculating the future value V_{an} of an annuity

$$V_{an} = A\frac{[(1+r)^n - 1]}{r} \qquad \text{from equation (2).}$$

Then $$P_{an} = \frac{V_{an}}{(1+r)^n} \qquad \text{from equation (3).}$$

Therefore:

$$P_{an} = \frac{A[(1+r)^n - 1]}{r(1+r)^n} \qquad (4).$$

Again the calculation may be simplified by using Table IV at the end of the book, which gives the present values of an annuity of £1 for different periods of years and different rates of interest. So if we expect a particular machine to produce a constant annual margin over costs of £6 for the next 5 years and if the interest rate is 12 per cent, the present value of the return from this investment is:

$$P_{an} = \frac{£6[(1 \cdot 12)^5 - 1]}{0 \cdot 12(1 \cdot 12)^5} = £21 \cdot 63.$$

Alternatively from Table IV we obtain:

$$P_{an} = £6 \times 3 \cdot 6048 = £21 \cdot 63.$$

We may use Tables III and IV together to convert a stream of irregular cash flows, such as may be obtained from a tree crop, into an equivalent annuity. Table III (or equation (3)) is first used to arrive at a present value for the stream of irregular cash flows. Then the present value is divided by the relevant value from Table IV to give a value for A, which is the equivalent annuity. Alternatively it can be calculated from equation (4) which when rearranged gives:

$$A = \frac{P_{an} \, r(1 + r)^n}{(1 + r)^n - 1}.$$

For example, if we have a project which is expected to yield an irregular series of cash flows over 20 years, which at 10 per cent interest gives a net present value of £14·60, the equivalent annual return is given by:

$$A = \frac{£14 \cdot 60 \times 0 \cdot 1 \, (1 \cdot 1)^{20}}{(1 \cdot 1)^{20} - 1} = £1 \cdot 72$$

or from Table IV

$$A = \frac{£14 \cdot 60}{8 \cdot 5136} = £1 \cdot 72.$$

This figure may then be used for comparison with alternative activities such as annual crop production which yield a regular annual return.

The equivalent annuity is also useful in deciding the optimum life for an investment. Thus if the extra return (actual, not discounted) obtainable by continuing the project a further year is greater than the equivalent annuity then it is profitable to continue the project. Alternatively if the extra return is less than the equivalent annuity, then total return would be maximized by abandoning the existing project and investing in a new one which presumably will yield the same equivalent annuity each year.

LAND VALUES

If the annuity is expected to continue into the indefinite future as is the case with land rents then $(1 + r)^n$ becomes very large so as n approaches infinity $(1 + r)^n - 1$ approaches $(1 + r)^n$. Thus when n is infinity:

$$P_{an} = \frac{A[(1 + r)^n - 1]}{r(1 + r)^n}$$

becomes

$$P_a = \frac{A(1 + r)^n}{r(1 + r)^n} = \frac{A}{r}. \tag{5}$$

Hence the present value or market price per hectare of land equals the annual rent A divided by the rate of interest r. This implies that rent represents the return on the capitalized sale value of the land. For example if the annual rent is £10 per hectare and the rate of interest 8 per cent, the present sale value per hectare would be $P_a = £10/0.08 = £125$.

We can now prove equation (5) in another way by assuming that a capital asset such as land with an infinite life will not depreciate. This means that its value will be the same at the end of a finite period of years as it was at the beginning, provided prices for other inputs and outputs do not change. The 'salvage value' of the asset at the end of n years is the same as the present value. Thus we may write:

$$P_a = \frac{A[(1 + r)^n - 1]}{r(1 + r)^n} + \frac{P_a}{(1 + r)^n}.$$

Subtracting $\dfrac{P_a}{(1 + r)^n}$ from both sides, then multiplying both sides by $(1 + r)^n$, gives:

$$P_a (1 + r)^n - P_a = \frac{A}{r}[(1 + r)^n - 1],$$

that is:

$$P_a \left[(1 + r)^n - 1 \right] = \frac{A}{r} \left[(1 + r)^n - 1 \right].$$

Dividing both sides by $\left[(1 + r)^n - 1 \right]$ leaves:

$$P_a = \frac{A}{r}.$$

THE INTERNAL RATE OF RETURN

So far we have used compounding and discounting to allow for the cost of capital, whether it is measured as an explicit rate of interest or an implicit cost arising from the farmer's personal time preference. Alternatively we can ignore the cost of waiting and use discounting (or compounding) to estimate the rate of return on capital instead. For this purpose, we find the 'rate of interest', which makes the present value of future benefits exactly equal to the present value of future costs, so that the net present value of the whole investment is zero.[9] The 'rate of interest' then measures the rate of return on capital, and is known as the 'internal rate of return' (IRR), the 'yield' of the investment or the 'marginal efficiency of capital'.

To return to the example of the timber plantation costing £30 and yielding £300 20 years later, we have a net present value given by:

$$\text{NPV} = \frac{£300}{(1 + r)^{20}} - £30.$$

If NPV is zero then

$$\frac{300}{(1 + r)^{20}} = 30,$$

therefore

$$(1 + r)^{20} = \frac{300}{30} = 10.$$

Hence $1 + r = 1 \cdot 122$ and $r = 0 \cdot 122$ or $12 \cdot 2$ per cent.

[9] The same result is obtained by compounding and finding the rate of interest which makes the future value of all cash flows at the end of the project life equal to zero.

Thus the internal rate of return is 12·2 per cent. At this rate of interest, £30 compounded is worth £300 at the end of 20 years *or* £300 in 20 years' time is worth £30 now when discounted.

In all but the simplest of cases, such as the above example, the internal rate of return cannot be calculated easily. It is therefore found by calculating net present values for a range of different interest rates, then plotting the results on a graph with interest rate on the horizontal axis and NPV on the vertical axis as shown

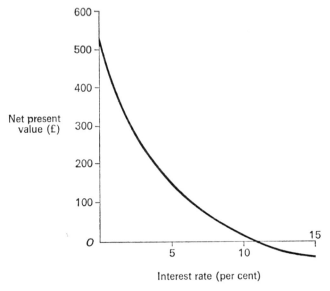

FIGURE 8.3 Estimating the internal rate of return. The interest rate at which net present value is zero, in this case 10·6 per cent, is the internal rate of return.

in Figure 8.3. The point where the graph cuts the rate of interest axis is the internal rate of return.

There is some controversy as to whether the net present value or the internal rate of return is the best criterion for assessing investments; in particular certain objections have been raised against the internal rate of return.[10] However, these objections are rarely of any practical importance and both methods are satisfactory. The choice of method then depends upon the main purpose of the discounting exercise. If the purpose is to allow for the cost of capital in estimating the return to some other resource, such as land,

---

[10] See for instance Hawkins, C. J. and Pearce, D. W. (1971) *Capital investment appraisal*, London, Macmillan, ch. 3.

then the net present value is the appropriate estimate. If, on the other hand, the purpose is to calculate the return on capital having estimated all the other costs of production, then the internal rate of return is more appropriate.

THE OIL PALM EXAMPLE

We may now return to our more realistic example of the oil palm plantation, for which estimated cash flows were given in Table 8.1 (page 155). To arrive at the net present value of these cash flows they all be discounted to year zero; that is to a point in time before any investment is made.[11] In this case the appropriate rate of interest is 8 per cent, or 0·08 as this is the rate charged on long-term loans by the Nigerian Credit Corporation.

The negative margin of £92·50 in year (1) is discounted 1 year to give:

$$P = \frac{-£92·50}{1·08} = -£85·60.$$

Likewise the negative cash flow for year (2) is discounted 2 years to give

$$P = \frac{-£19·15}{(1·08)^2} = -£16·40.$$

Similarly the cash flow for each successive year is discounted to year zero. For the final year (35), the present value is

$$P = \frac{+£22·58}{(1·08)^{35}} = +£1·53.$$

The sum of all these discounted cash flows then equals the net present value for the whole life of the investment. It amounts to £48·40 per hectare of oil palms.

The equivalent annuity is calculated to be:

$$A = \frac{£48·40 \times 0·08 \, (1·08)^{35}}{(1·08)^{35} - 1} = £4·15 \text{ per hectare per year.}$$

The internal rate of return is determined by calculating the net present value at various rates of interest from 0 up to 16 per cent. The graph is plotted in Figure 8.3. From this it is clear that the internal rate of return is 10·6 per cent.

[11] See Upton, M. (1966) Tree crops: a long-term investment, *Journal of Agricultural Economics*, vol. 17, no. 1, p. 82.

It should be noted that if the revenue could be increased in the early years of the investment, say by growing arable crops between the young trees, both the net present value and the internal rate of return would be increased.

The value of the trees can be estimated at any stage of their life either in terms of the total cost of establishment including compound interest, or in terms of the discounted value of expected future returns.[12] The former represents the cost and the latter the expected benefit, so clearly for any profitable investment the discounted value of future returns will exceed the cumulative capital

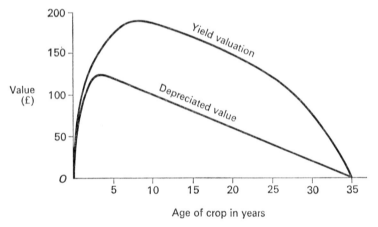

FIGURE 8.4 Valuations of one hectare of oil palms. The yield valuation is greater than the value estimated by the traditional depreciated cost method throughout the life of the crop.

cost. Thus at an interest rate of 10 per cent the total capital cost of a hectare of oil palms 20 years old is the first year cash flow of —£92·50 compounded 19 years, plus the second year cash flow of —£19·15 compounded 18 years and so on. The sum of all these cash flows plus the interest at 10 per cent accumulated amounts to a total cost of £114·50.

The present value of future cash flows from year (20) onwards will be the margin in year (21) of +£22·58 discounted 1 year, plus the margin in year (22) discounted 2 years and so on up to year (35). The present value in year (20) of all these future benefits amounts to £171·20.

[12] See Upton, M. (1969) Note on tree crops: the capital profile and valuations, *Journal of Agricultural Economics*, vol. 20, no. 1 p. 143.

If the internal rate of return is used instead of the rate of interest for compounding or discounting, the valuation is the same no matter which method of calculation is used. Thus the value of the hectare of oil palms at the end of 20 years at the internal rate of return or yield of 10·6 per cent is £165·70. This we may call the 'yield valuation'. The pattern of yield valuations over the life of the hectare of oil palms is given in Figure 8.4, together with the valuations estimated by simple depreciation.

9 Management

THE PRODUCTIVITY OF MANAGEMENT

On family farms the functions of management are provided by the farmer himself. In this he may be helped by other members of his family, by professional advisers or even by friends and neighbours. Nevertheless he alone must see that the decisions are carried out and must bear the consequences; he takes the risks. A farmer, in this context, means any person who controls some agricultural resources and makes decisions about their use. Thus on family farms, wives or grown sons who cultivate some plots for themselves in addition to providing labour for the main farm also provide the management on their own plots.

It is very difficult, if not impossible, to measure the quantity of management available or in use. Clearly the number of man-hours spent on managerial activity is not a very meaningful guide because the quality of management decisions is all important. One man may spend much longer than his neighbour thinking and planning the future, but it does not necessarily follow that his decisions will be any different from his neighbour's. Nevertheless, despite the difficulty of measuring the amount in use, management is plainly a productive resource.[1]

Physical resources of land, labour and capital are not productive unless they are organized and coordinated by someone who makes the necessary decisions and carries them out. In fact, without inputs of management a farm would not exist. Hence profit, the difference between the total output and the costs of physical inputs, is really the product of management. Of course, management is not productive on its own, it must have other resources to work with, but this is also true of land, labour or capital.

Even within a single village or type of farming area there are wide differences between farmers in the amounts they produce

---

[1] For a contrary view see Johnson, G. L. (1964) A note on non-conventional inputs and conventional production functions, *in* Eicher, C. and Witt, L. (eds.) *Agriculture in economic development*, New York, McGraw-Hill.

and the profits they make. These differences can only partially be explained by differences in the quality and quantity of land, labour and capital. Indeed in situations, as on some farm settlement schemes, where farmers are all allocated similar quantities of land and other resources under similar environmental conditions and where they are expected to grow the same combination of crops, there are still quite large differences in the outputs produced and profits earned. Some of the differences may be accounted for simply in terms of luck. One farmer falls ill at a critical period, another has his crops attacked by locusts, another faces a drought just after he has finished planting so his crop fails. The farmer who escapes such disasters will produce more output and make more profit than his less fortunate neighbours. However, over a period of several years it is unlikely that one farmer will be lucky all the time, while his neighbours are unlucky all the time. The farmer who seems to be lucky is probably more thorough in caring for his crops and livestock, and takes more precautions against such disasters than his 'unlucky' neighbour. Thus much of the variation in farmers' incomes and profits must be explained by variation in inputs of management.

Where farms vary in size much of the variation may result from differences in management. If, as a result of differences in management, the marginal product per unit of labour is higher than average on a particular farm, there is an incentive for labour to move onto that farm (see Figure 9.1). In this way differences in management cause differences in the economic optimum level of resource use. In fact, as we saw in Chapter 3, the attitudes and ability of the farmer as a manager set a limit on the size of farm he is willing or able to manage. Variations in farmers' attitudes and abilities therefore cause variations in farm sizes. This effect is quite likely to be more marked where farmers use credit to expand their farms. Credit institutions, both private and government sponsored, will be more willing to lend to those farmers they judge to be 'good' managers, because the risks of project failure and non-repayment of the loan are less. As a result the more able managers find it easier to raise loans to expand their farms.

Finally differences between farmers in their attitudes and ability cause differences in the rate at which they adopt profitable innovations. While a few farmers adopt an innovation rapidly, the bulk of producers take more time to become aware of the new practice or idea and to evaluate its benefits. If over a range of different innovations a farmer is persistently among the first to adopt the new product or practice he can be classified as an innovator, and so on through the time scale to the other extreme, where, if he is consistently slow in adopting each innovation, he is classified as a

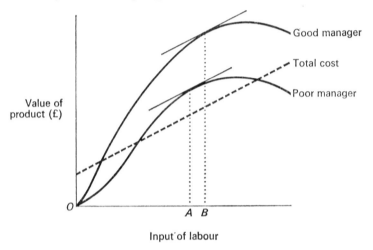

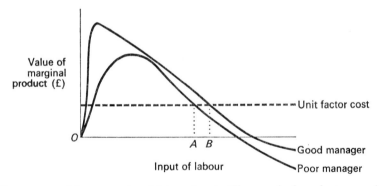

FIGURE 9.1 Management and labour inputs. The marginal product per unit of a given level of labour input is increased by better management. Hence the economic optimum level of labour input is increased from *A* to *B*. Compare with Figure 2.6.

laggard.[2] Now provided that the innovations are profitable, that is they either increase output or decrease costs, innovators are likely to make more profit than other farmers and laggards are likely to make less. In fact some economists consider that profit is the reward for innovation.

[2] See Rogers, E. M. (1962) *The diffusion of innovations*, Glencoe, Illinois, Free Press.

The personal characteristics of the farmer which directly influence the decisions he makes can be summed up under the broad headings of attitudes and objectives on the one hand and managerial ability on the other. These characteristics are dependent, in turn, on the farmer's background and environment. The relationships are summarized in diagram form in Figure 9.2.

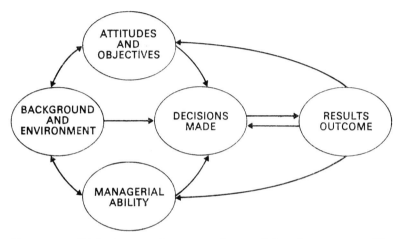

FIGURE 9.2 Links between important characteristics of a manager. The arrows pointing back suggest that results obtained currently may influence attitudes and objectives, managerial ability and decisions made in the future.

Farmers' backgrounds, attitudes and abilities must all be taken into account in farm management analysis and planning. Plans which do not take account of the farmer as a person are likely to fail.

ATTITUDES AND OBJECTIVES

It is widely assumed that the traditionalism of peasant farmers is a major obstacle to agricultural development. Rural culture is contrasted with the city or urban culture where no long-established pattern of behaviour exists and change is readily accepted. The characteristics of rural, peasant or folk cultures include personal relationships based on status rather than contract, fatalistic outlook and limited horizons and aspirations.

Personal relationships in small village communities are based on the relative status of the individuals concerned. Almost everyone living in the village was born and raised there so his social position is well established. He deals only with his neighbours or kinsmen and may mistrust outsiders. Impersonal contractual relationships between comparative strangers are practically unknown. Thus the quantity of resources a farmer controls depends upon his status in the village society rather than upon his ability to manage them. Indeed, there may be pressures on the individual to conform to locally accepted standards of living and methods of production. A farmer who invests much more than his neighbours in new forms of capital may face public disapproval and perhaps withdrawal of cooperation by his kinsfolk and neighbours. Furthermore an individual who is successful in any kind of productive activity is expected to share the fruits of his success with his kinsfolk.

In such circumstances the safest way for a successful farmer to spend his money is in building up his prestige and status rather than in directly productive investment. Lavish generosity is expected; the man who saves to invest is likely to be branded a miser. Wealth is often interpreted not in terms of obvious ownership of resources or a higher standard of living but in terms of the number of people against whom one has a claim. The wealthy men do become leaders, not only within their descent group but in the community at large.[3]

The fatalistic outlook of rural people means that they lack confidence in their own ability to influence or control events. Success is thought to be due to the assistance of the gods or spirits and failure to their disfavour. Because the farmers' perception of the possibility of self-help is low, dependence upon government is high. Many farmers seem to take the view that conditions in their villages will only improve if the government helps. Linked with the fatalistic outlook is a desire to avoid risk and an unwillingness to innovate. If the farmer views agriculture as a struggle against supernatural forces which are stronger than himself then his best policy is to be prepared for the worst and to stick to reliable and well-tried practices. However, as we saw in Chapter 5, risk avoidance may simply reflect the poverty of small farmers in that they cannot afford to take risks.

Limited horizons and aspirations may be related to lack of communications. In remote areas farmers really have no opportunity to learn of the world outside the village and the consumer goods available there. Thus the traditional farmer's wants are

[3] See Lloyd, P. C. (1967) *Africa in social change*, London, Penguin Books; and Colson, E. (1959) Native cultural and social patterns in contemporary Africa, *in* Haines, C. G. (ed.) *Africa today*, Baltimore, Johns Hopkins Press.

limited to little more than the subsistence needs of his family. He has no incentive to increase production. At the same time he is supposed to have limited time horizons. He does not prepare for the future but has a strong preference for current consumption, so the cost of saving and investment is high.

It should be noted that all these assumed characteristics of rural people may or may not be found in practice. Field studies have been carried out in some parts of Africa but the results are by no means conclusive. Furthermore these characteristics need not form a major obstacle to economic development. Thus if personal relationships are based on status rather than contract a man may increase the size of his farm business indirectly by first spending to increase his status and then using his increased status to expand his farm. In any case, as we have seen in discussing land, labour and capital, where opportunities have occurred for agricultural expansion, particularly in major cash-cropping areas, cultural changes have occurred and land is sold or leased, labour is hired and capital borrowed on a contractual basis. Similarly outlook, horizons and aspirations are all likely to change quite rapidly if communications and opportunities improve.

Within most communities today, already involved in the processes of change, there are wide differences between individuals in their attitudes and objectives. Some farmers are more 'urbanized' than others in their outlook.[4] Thus some are less bound by convention and the need to conform to accepted patterns of behaviour than others; some are more rational, less fatalistic and more willing to take risks and innovate; some have higher aspirations, greater need for achievement and are more willing to save for the future. These differences are important in understanding differences in farm incomes and in planning agricultural development. It is important to identify which farmers are more urbanized in outlook and more willing to innovate when attempting to introduce new techniques and products.

MANAGERIAL ABILITY

Farm management consists of taking and carrying out decisions and farmers differ considerably in their ability to do this. Farm management decisions vary widely in their scope and importance. At one extreme there is the major decision of what system of farming to follow; which crops to produce, how many hectares of each

---

[4] See Upton, *Agriculture in south-western Nigeria*, part 3.

and how many livestock to keep. Such major decisions may be taken at relatively wide intervals, perhaps once a year, perhaps at longer intervals. Usually the farmer will want to spend some time thinking about the problem before reaching a decision and he may ask the help and advice of others. At the other extreme, there are minor decisions which arise all the time, such as whether a particular crop is ready for harvest, should an extra hour be spent in weeding, is it worth keeping a sick hen or should it be killed. These decisions must often be made immediately, on the spot. Some farm management advisers think that a farmer may be good at taking the major planning decisions and bad at the day-to-day management or conversely bad at the first and good at the second. However, in practice it is uncertain whether a clear distinction can be made. The farm system may have developed gradually as the result of a series of minor decisions rather than of a single broad plan.

Another distinction which is sometimes made in considering the overall efficiency of management is that between technical efficiency and allocative or economic efficiency. Technical efficiency is measured by relating the level of output achieved to the quantities of inputs used. A manager who is technically efficient produces more than average output from a given combination of resources. He gets high yields. Economic efficiency on the other hand refers to the choice and combination of resources and enterprises. The manager who is economically efficient uses his resources at or near the economic optimum. The distinction is illustrated in Figures 9.3 and 9.4. Again this distinction is not very clear in practice. All the decisions which influence the level of output obtained (technical efficiency) are really economic decisions. In practice, particularly in the African context, the quantity and combination of resources available may be quite outside the farmer's control. The area of land available, the family labour force or the small amount of capital may be quite fixed in supply.

While decision making is often considered the central or key task of the manager, the whole process of reaching a decision and putting it into practice can be broken down into the following steps:

(1) recognition of a problem;
(2) observation or collection of relevant facts;
(3) analysis and specification of alternatives;
(4) choice of alternatives (decision making or planning);
(5) taking action (putting plan into practice).

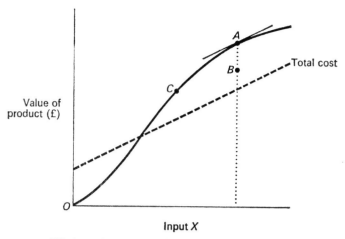

FIGURE 9.3 Efficiency in factor-product relationships. If point *A* is the economic optimum then the farmer operating at point *B* is said to be technically inefficient and the farmer operating at point *C* is said to be economically inefficient.

These steps are not independent and they may not always occur in the same order. For instance some observation of relevant facts might follow the analysis and specification of alternatives (see later chapters on farm planning).

However, this simple analysis of the management process suggests which personal characteristics are likely to be associated with managerial ability. Obviously knowledge is one such characteristic. Knowledge of a considerable number of different methods of production (or of marketing) enables a farmer to compare his present methods with possible alternatives, to recognize problems of low output or high inputs and to collect relevant facts. A farmer who only knows one traditional method of production cannot appreciate its defects or assess the possible benefits of a suggested change. Closely associated with knowledge are the skills necessary for decision making and for putting decisions into action. These include the skills of farm management analysis and planning and the practical skills involved in agricultural production. To some extent these skills can be learned, indeed this book is aimed at teaching the skills of farm management analysis and planning, but there is no doubt some variation in inherent ability. Openness of mind and ability to manipulate ideas are useful characteristics in recognizing problems, observing facts and analysing and specifying alternatives.

Finally the manager needs the necessary motivation and drive to put the decision into practice. This characteristic serves to distinguish the 'practical man' from the less practical. Some individuals seem capable of carrying out all steps up to and including the actual decision making but then just cannot put them into practice successfully. Where the farmer provides his own labour force, this simply requires him to assemble the necessary materials and equipment and to possess the desire or motivation to get the

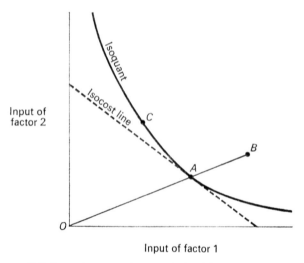

FIGURE 9.4 Efficiency in factor-factor relationships. If point *A* is the least-cost combination for a given level of output then a farmer using the combination of inputs represented by point *B* (to produce the same level of output) is technically inefficient and one who uses the combination at point *C* is economically inefficient.

job done. Where there are others involved in the labour force he must motivate and control these others. Ability to organize and lead other people is then needed.

Managerial ability must be taken into account in planning agricultural change. Some farmers may be incapable of managing larger and more complex methods and systems of farming. Thus where irrigation or tractor cultivation is introduced into an area where experience of this form of agriculture is lacking, it may be necessary to provide management from outside in the initial stages. The cultivators are then little more than hired workers under someone else's supervision. However this form of organization is

costly and wasteful of the potential managerial ability of the cultivators so it is desirable that responsibility and control should
eventually pass into the hands of the farmers themselves.

INFLUENCE OF ENVIRONMENT AND BACKGROUND

The environment, both physical and social, has an important influence on all these human factors. For instance a farmer's attitudes and goals are closely related with those of the community as
a whole. Frequently there is a wish to conform to locally accepted
standards of behaviour, in farming as in other activities. This may
prevent farmers from adopting new crops and techniques or from
expanding their scale of farming far beyond the local average.
Alternatively, in societies where profit making and material
achievement are respected and admired, then a man may be
driven to greater efforts to impress his neighbours. Where the
physical environment makes agriculture difficult, as in the semi-
arid zones, and where the fear of natural catastrophe is always
present, farmers will be unwilling to take the risk of trying new
ideas and new techniques of production. Where the technology of
agriculture has begun to change, however, and people have begun
to enjoy increased production through the adoption of new practices, the social values shift towards attaching higher value to individual experimentation and the introduction of still more new
methods.

Education has a very important influence on farmers' attitudes
and abilities. Used in its broadest sense education must include
not only formal schooling but also agricultural extension and indeed any experience which broadens a man's outlook and increases his knowledge. Some farmers improve their education by
travel. They visit other agricultural regions or farms and gain new
knowledge and new ideas as a consequence. The better educated
farmer not only has more knowledge of alternative enterprises and
techniques than his uneducated neighbour but also has more confidence in his own judgement and feels less need for the approval of
others. Thus it is frequently found that better educated farmers are
more concerned with profit making, more willing to take risks and
better able to take decisions than those with less education. On the
other hand education may evoke a desire for leisure, which may
lead to shorter working days. In many countries formal education
is seen as a way to move out of agriculture and not as a method of
increasing efficiency in farming.

The farmer's age also has an influence on management perform-ance although the overall direction of this influence is not clear. On the one hand as a man ages he gains experience and we would expect his decision-making ability to improve. On the other hand it is generally found that goals change with increasing age, usually towards increasing leisure and reducing work. The effect of age is further confused by the stage reached in the family life cycle. The farmer makes his decisions in the light of his membership of the family group and the influence which other members of the family have on him. Thus a farmer with a young family dependent on him has more incentive to maximize his cash profit than a man with few dependents. At the same time sons may have a consider-able influence on their fathers, perhaps in persuading them to adopt risky innovations they would not otherwise consider.

SUGGESTIONS FOR FURTHER READING

BENVENUTI, B. (1962) *Farming in cultural change*, Assen, Van Gorcum.

BOWDEN, E. and MORRIS, J. (1969) Social characteristics of progressive Baganda farmers, *East African Journal of Rural Development*, vol. 2, no. 1, p. 56.

JONES, G. E. (1967) The adoption and diffusion of agricultural practices, *World Agricultural Economics and Rural Sociology Abstracts*, vol. 9, no. 3, p. 1.

LEVER, B. G. (1970) *Agricultural extension in Botswana*, University of Reading, Department of Agricultural Economics, Development Study no. 7.

MUGGEN, G. (1969) Human factors and farm management, a review of the literature, *World Agricultural Economics and Rural Sociology Abstracts*, vol. 11, no. 2, p. 1.

ROGERS, E. M. (1962) *The diffusion of innovations*, Glencoe, Illinois, Free Press.

UPTON, M. (1967) *Agriculture in south-western Nigeria*, University of Reading, Department of Agricultural Economics, Development Study no. 3.

10 Technical experiments and case studies

FARM MANAGEMENT DATA AND THEIR USES

Farm management investigations are aimed at providing data for any of the following purposes.

To describe existing farming systems and to estimate inputs and outputs, costs and returns. An understanding of existing farming systems is necessary for planning both national and regional agricultural policies. For instance production data are needed when choosing between alternative systems of land tenure, for fixing the price of cash crops, for estimating the benefits from agricultural credit schemes and for deciding on levels of agricultural taxation. Likewise an understanding of existing farming systems is necessary to suggest directions for future technical and economic research. The benefits of the research are likely to be greater if it is directed towards major crops rather than minor ones, towards saving labour at busy times rather than slack times, towards changes which will be socially acceptable rather than those which may be unacceptable.

For planning changes in existing systems. Farm planning data include not only information on the existing systems but also estimates of the effects of introducing new techniques and products. Although in areas of traditional agriculture there may be few alternatives and little scope for improvement without introducing innovations from outside, where change is already occurring there may be scope for improvement simply by reallocating existing resources. In order to plan or analyse the effects of reallocating resources standard production data are needed. For an extension worker to judge whether crops or livestock are yielding as they ought, he needs production standards on which to base his judgement. Thus data are needed for planning and estimating the likely effects of changes *before* the changes are made.

For measuring the effects of changes after they have been made. Investigations may be aimed at comparing results obtained in practice from different alternative products and methods of production.

Data for these purposes may be obtained from three main sources, namely technical experiments, case-study (unit) farms or farm surveys. A fourth category of personal experience might be added although the experience must have been gained from one of the first three sources in the past. Some would argue that farm surveys are the most appropriate method of providing descriptive data on existing farming systems. The survey results may suggest ideas for technical experiments which would provide planning data. The most attractive plans might then be tried out in practice on case-study farms before being recommended to farmers by the extension service. However, it should not be assumed that each source is suited to the provision of only one kind of data. The choice between technical experiments, case studies or surveys depends upon several considerations not least of which is cost.

THE DATA NEEDED

For some purposes the investigation may only be concerned with a single enterprise. Thus a study intended to provide a base for fixing the price of coffee might be restricted to this crop only. Such 'enterprise cost studies' may be useful in areas where only one enterprise is of major economic importance, and where it might be difficult to obtain information on the whole farm.

However, there are dangers in studying one enterprise in isolation since important supplementary and complementary relationships between enterprises may be missed. Without considering the whole farm it is difficult to assess the opportunity costs of the resources used.

Whether the investigation is concerned with a single enterprise or the whole farm the information needed can be listed in four broad groups:

1. *Descriptive material on the pattern of farming*. This will include information on what is produced and the methods of production.

2. *Input–output data in physical terms*. These include quantities of resources used and physical product obtained. Inputs and outputs must be measured over a specific period of time, usually a year. This is perfectly satisfactory for measuring inputs of seed and manure and output for an annual crop but may raise problems

with longer term investments such as tree crops, breeding livestock, buildings and machinery. Where much of the output is
consumed by the farm family it is necessary to record products
consumed as well as those sold.

3. *Value placed on inputs and outputs.* For reasons already given
market prices are generally used as a means of evaluating inputs
and outputs. For certain purposes, other measures of value may be
used. Thus data for planning famine relief might be based upon
nutritional measures of value. In some cases it may be considered
necessary to investigate farmers' attitudes and objectives in order
to evaluate inputs and outputs in their terms. Alternatively where
prices are fixed by government these may be the appropriate
values to use.

4. *Economic and social constraints.* Economic constraints are actual
physical limitations on certain resource supplies. For instance the
labour force may be restricted to family members. It is then necessary to assess what labour supply is available on average. The same
may be true of land and for most forms of capital.

Social constraints may result from local customs, taboos or mere
preferences which may preclude the production of certain kinds of
livestock or crop products or the use of credit or draft livestock. In
so far as such constraints influence the pattern of production, they
should be included as some of the necessary farm management
data. In order to investigate the dynamics of the system it may also
be necessary to evaluate consumption, saving and population
growth.

EXISTING SOURCES

The first source of data which any farm management adviser
should draw upon is that of local experience. In particular, officers
of the agricultural extension service can provide information on
local farming systems and estimates of inputs and outputs. Information from individuals is both cheaply and quickly acquired.
Where results are needed rapidly, this may be the only source of
data which it is possible to use. Many studies have been based on
local estimates of inputs and outputs. For example, the annual
costs and returns for oil palms quoted in Table 8.1 were derived
in this way. *An agricultural notebook* by Phillips is a collection of
estimates of inputs and outputs for the major crops of Nigeria.[1]

Clearly the reliability of such estimates must depend upon their

[1] Phillips, T. A. (1964) *An agricultural notebook*, 2nd ed., Ikeja, Longmans of Nigeria.

original source. If they are based on surveys and experiments they may be accurate but nevertheless could be misleading if the original data were collected under different conditions, many years earlier or in another locality. If however they are merely impressions gained from working in the area the errors could be large. There is probably a tendency for agricultural officers to be over-optimistic about the agricultural potential of their particular areas.

A second source of data is published material. During the past sixty years a great deal of agricultural research has been done in Africa, much of it on government experimental stations. Only a relatively small part of the research done has been published but this may provide information on agricultural inputs and outputs. Here again the data may not be directly applicable to the farms currently under investigation. However the difference between regions and between time periods may not be significant when compared with variations between farms within one area. In particular the average annual labour inputs per acre of a given crop may not vary significantly from one country to another or from one time period to another.

Farm management data may also be extracted from field surveys, perhaps the best established form of which is the agricultural census. Censuses are carried out mainly by government statistics departments, although veterinary and agricultural services may be involved and international bodies sometimes provide both financial support and personnel. The main types of information gathered concern crop areas, holding size, crop yields and livestock counts. These sorts of data were collected in some countries by agricultural and veterinary staff using methods of dubious validity up until the 1950s. In most countries of Africa, the first formal agricultural census was taken around 1950 in connection with the 1950 World Agricultural Census. This led to an increasing realization of the value of the accurate measurement of such characteristics of the farm as size, form of tenure, utilization of area and agricultural equipment inventories. Characteristics of the human and livestock population and value of crop and livestock products and sometimes the type of purchased inputs used were also recorded. Such a comprehensive framework was not used, however, until the F.A.O. instigated the 1960 World Census of Agriculture. Generally, it is proposed to continue making these studies at ten-year intervals.

Agricultural censuses usually only provide general descriptive material for heading (1) above and information on the resources available applicable to heading (4) above. These studies do not give input–output relationships or price data. In some countries, market statistics are collected, particularly the quantities and prices of export crops, though prices of local food crops may also be recorded.

Thus it is probable that descriptive material on the local pattern of farming is already available. There may also be information on local prices and on the economic constraints, in terms of the resources available in the region. However, there are rarely sufficient data on the physical relationships between inputs and outputs. Where these are lacking, research must be carried out to provide original figures on inputs and outputs.

TECHNICAL EXPERIMENTS

Technical experiments, carried out on research stations or even on farm trials, give precise and accurate data on the effects of varying certain inputs on a particular crop or animal under certain specific conditions. In comparison with actual farm studies, experiments raise comparatively few organizational problems and probably represent the simplest and cheapest method of collecting input–output data. This method of research is the obvious choice for testing new technology and measuring the productivity of new inputs which have not yet been introduced on farms. However, experiments cannot provide any information on existing patterns of farming, local prices, or social and economic constraints.

Some would argue that technical experimentation is the responsibility of agricultural scientists and technologists and not of management specialists or economists. This view cannot be justified. The economic approach is relevant both in suggesting topics for research and guiding the way in which experiments are carried out and analysed.

The need for effective links between economists and technical research workers is emphasized in a study of experiences with agricultural development in tropical Africa undertaken for the World Bank.[2] The authors suggest 'that to a considerable extent there has been a failure to develop new agricultural inputs which are really rewarding to the farmer, that research has been poorly tailored to actual development needs and has not effectively linked

[2] de Wilde, J. C. and others, *Experiences with agricultural development in tropical Africa*: vol. 1, *The synthesis*.

economics and technology.' They suggest that past weaknesses arose basically from a lack of economic understanding and appreciation of existing methods of agriculture. It is suggested that this may also be true of indigenous scientists who have been trained overseas. The choice of topics for experimentation often reflects a technological bias, particularly in an exclusive preoccupation with increasing outputs per unit of area without regard to all the related inputs that are necessary and the farmer's capacity to provide them. Thus in many parts of Africa seasonal labour peaks represent a more critical constraint than land. Interaction between enterprises is often ignored and the opportunity costs of resources wrongly assessed.

To improve the situation the authors of this study suggest that an agricultural economist should be assigned to each of the more important research stations or at least should spend some time at research stations to evaluate the economics of research recommendations and improve the economic orientation of new and current research. These suggestions are highly convincing as is their recommendation that results of experiments should be tested on a pilot scale or on case-study farms before being put into general practice or recommended to farmers.

Farm trials are usually necessary because both labour inputs and product outputs are generally much higher under experimental conditions than they are on farms. Naturally the standard of management of small experimental plots on government research stations is much better than that on privately owned farms. For this reason, technical experiments are likely to give a biased picture of the inputs and outputs likely to be found in practice. A further difficulty here is that most research work concerns pure crop stands, while farmers almost invariably practice mixed cropping.

CASE STUDIES AND UNIT FARMS

As an alternative to building up estimates of the costs and returns in farming from existing sources, and the results of technical experiments, we may collect information from farmers themselves. A 'case study' is an investigation of a single representative farm, whereas a 'survey' is an investigation of a number of representative farms, known as a sample of farms.

On large, commercial farms where records and accounts are kept for management purposes, each farm may be treated as an individual case study. The records and accounts are used to analyse weaknesses in the existing system and to provide planning data for the individual farm.[3] However a comparison of accounts on different farms in the same environment often provides useful information on possible ways of increasing individual farm incomes. Where farmers are not able or inclined to keep accounts themselves, information required by farm management advisers must be collected from the farmers by field recorders.

Case studies can provide information on the pattern and methods of production, the prices of inputs and outputs and on social and economic constraints on the existing levels of output. The errors of measurement or estimation of inputs and outputs are likely to be larger on actual farms than on closely controlled experiments, and with a case-study farm, unlike an experiment, there is no way of testing the reliability of the results obtained.

Since with a survey each item of information is collected from a large number of farms, the cost is likely to be higher than for a case study. Alternatively, for a given cost and effort, much more detailed information can be obtained from a case study than from a survey. However, to set against this there is the problem of selecting the farm for the case study. For the data to be applicable on other farms of a given type or area, it is essential that the one chosen for case study is representative or typical of the others: and it is difficult to ensure this. When a farm *is* selected, there is no way of estimating the variation likely to be found between other farms of the general type being studied.

If a case-study farm is chosen at random, or solely on the basis of the farmer's willingness to cooperate, there is a strong chance that it will differ considerably from most farms of that general type and area. In fact a conscious effort must be made to find a typical farm, but this means that the researcher needs a previous knowledge of farms of that type in order to identify a typical one. This may imply that if the knowledge is lacking, some kind of survey must be carried out first, in order to define a typical farm. Even then, it may be difficult to find one typical in all respects. One farmer, with a family labour force similar in size to most farms in the district being studied, may keep an unusually large number of poultry; another farmer, typical in both these respects, may be unusual in using a tractor hire service.

Clayton has argued that these problems of selecting a representa-

[3] Methods of keeping records and accounts are discussed in Upton, M. and Anthonio, Q. B. O. (1965) *Farming as a business*, Oxford University Press, chapters 8 and 9.

tive farm do not apply in Africa particularly where high value cash or export crops are important in farming systems.[4] He considers that production depends on the combination of only two factors, land and labour, so that in many areas it is possible to define a peasant holding fairly accurately by its land/labour ratio. It is difficult to support this view, since management decisions and some, albeit small, amounts of capital also influence the level of outputs, and vary between farmers.

Nevertheless, case studies have been used widely to provide farm management data for various types of farming in Africa.[5]

A variation of the case-study approach is the use of unit farms to test the reliability of farm plans from an economic and technical point of view. For this purpose the investigator must be able to exercise complete control over the farm on which he is experimenting. His control is not only in respect of agricultural techniques, but also in the more fundamental features of labour organization, capital investment, enterprise combination, level of resource use and so on. By proper accounting the investigator can collect all the facts about the operation of a given farm plan. He can control and examine any reorganization he may think desirable in order to achieve an economically more efficient farm. Since all resource inputs and outputs are recorded, together with market prices where relevant, unit farms may be useful sources of farm planning data, provided, of course, they approximate to typical farming situations.

Unit farms have been established at one time or another in practically every country in Africa, but very few reports have been published.

One notable exception is the unit farm at the Western Research Centre, Ukiriguru, Tanzania, for which the results from 1962–5 have been published by Collinson.[6] Conception of this unit, and many others in Africa, owed much to the work of Dr. A. L. Jolly in

[4] Clayton, E. S. (1965) *Agrarian development in peasant economies: some lessons from Kenya*, Oxford, Pergamon Press.

[5] See for instance Ruthenberg, H. (ed.) (1968) *Smallholder farming and smallholder development in Tanzania*, Munich, IFO Institut; Kenya Farm Economics Survey Unit, *Reports* nos. 13, 15, 18 and 23, Statistics Division, P.O. Box 30266, Nairobi; Hoffman, H. K. F. (1967) *Case studies of progressive farming in central Malawi*, Malawi Government; and Foster, P. and Yost, L. (1967) *Buganda rudimentary sedentary agriculture*, University of Maryland, Department of Agricultural Economics, Miscellaneous Publication no. 590.

[6] Collinson, M. P. (1970) Experience with a trial management farm in Tanzania, *East African Journal of Rural Development*, vol. 2, no. 2, p. 28.

Trinidad who wrote widely on the uses of this approach.[7] His arguments in favour of the unit farm approach are worth quoting at length:

The need for such well controlled farm management experiments is generally underestimated particularly in the tropics. Agricultural science is concerned essentially with the technical details of farming and not with the organization of them into an efficient farm business. The scientist cannot judge with any accuracy, nor for that matter is he greatly concerned, how important a new technique may be to the operation of the farm or how its introduction will affect the farm organization in other respects. To him, every isolated improvement is another brick in the building of a more efficient agriculture. In reality, however, every improved technique affects the whole structure of the farm. Its introduction does not represent the laying of a brick on top of a building, but the removal of one part way down and replacing it by a better one. This replacement can be as disturbing to a farm as to a building.

Wherever agriculture is technically less advanced the effect of any single innovation is likely to be even more fundamental and the need for subjecting it to economic tests within a farming system is even greater. The admirable work done by veterinary officers in the tropics has unfortunately led in some areas to actual deterioration in agriculture through overstocking. Animal health has been regarded rather as an object in itself and has been dissociated from the management of any particular type of farm. Similarly the introduction of a cash crop into a tropical subsistence farming district, although it may by itself represent a technical and economic advance, may result in a deterioration in the agriculture. The danger is over-concentration on the one crop to the destruction of a more balanced and therefore more permanent and flexible form of agriculture.

The primary function of the farm management economist should be to study the repercussions of technical innovation on farm organization: some economists, particularly those who are concerned with less developed areas, might go further to say that their function is to select combinations of enterprises and practices from the fund of scientific knowledge that will make a more efficient use of the economic resources available. Whatever the degree of function, the economist is generally faced with the difficulty that

[7] See for instance Jolly, A. L. (1952) Unit farms, *Tropical Agriculture*, Trinidad, vol. 29, no. 7, p. 12; Jolly, A. L. (1954) *Report on peasant experimental farms at the Imperial College of Tropical Agriculture*, Trinidad, Imperial College of Tropical Agriculture, Economic Series no. 2; Jolly, A. L. (1955) Peasant experimental farms, *Tropical Agriculture*, Trinidad, vol. 32, no. 4, p. 257; and Jolly, A. L. (1957) The unit farm as a tool in farm management research, *Journal of Farm Economics*, vol. 39, no. 3, p. 739.

he has no laboratory in which to carry out experiments in organization. The classical tools of economic research are designed essentially to measure conditions as they exist and changes as they occur.

Since Jolly was writing there has been a rapid development in techniques of farm planning which may have reduced the need for field trials of new management systems. Nevertheless his argument has much strength even today.

Both Jolly and Collinson suggest that in many cases the establishment of unit farms has been unsatisfactory because of confusion of objectives and the resultant faulty technique of investigation. Thus, as suggested above, the main objective is usually experimental: that is to try out new and improved techniques and farming systems in practice. Even here there is some debate as to whether the system should be tried with existing institutional factors, such as local forms of land tenure and credit, or whether the economic environment should be modified in the interests of greater efficiency. However, in some cases unit farms have been set up with the additional objectives of providing data under typical average management and of serving as a demonstration of good management to other farmers. These different objectives may require different methods of organization, hence the possible confusion when two or more objectives are followed at the same time.

The method of establishment of unit farms therefore varies from place to place. One possibility is simply to offer incentives to an independent farmer, by providing capital and technical advice. In return he would be expected to follow the system laid down and keep records. Experience shows that this approach is unlikely to succeed. Alternatively the unit must be established on the research station, where the activities of the farmer and his family can be more closely controlled and supervised. The farmer may work on a profit sharing basis, also sharing some of the decision making, or he and his family may be employed solely for wages to operate the system laid down by the farm management specialist. The last approach, of course, gives the researcher the highest degree of control but the situation is furthest removed from a real farm situation. Some share of profits may be necessary as an incentive to work.

The major disadvantage with the unit farm approach, as with any case study, is that since the results are obtained from a small number it is very dangerous to generalize from them. The particular farms chosen may be on very good soil or the farmers may work harder, be better managers or luckier than average. Of course they may be typical but there is no way of testing whether this is the case. Thus a system which is successful on a unit farm may still fail on the majority of farms in the district when it is introduced by the extension service. Frequently unit farms are used to compare contrasting systems, as where one unit is operated with ox draft and another nearby is operated with tractors. This kind of comparison is almost certain to be misleading since it is most unlikely that both farmers will be of identical managerial ability and technical skill. Over a short period one may be luckier than the other. Differences between the results obtained on the two units could be due to many factors other than the difference in farm systems.

Nevertheless, despite this obvious disadvantage, the unit farm approach is often the least costly and hence the most attractive method of assessing new techniques and patterns of production at the farm level.

SUGGESTIONS FOR FURTHER READING

BARNARD, C. S. (1963) Farm models, management objectives and the bounded planning environment, *Journal of Agricultural Economics*, vol. 15, no. 4, p. 525.

CARTER, H. C. (1963) Representative farms—guides for decision making, *Journal of Farm Economics*, vol. 45, no. 5, p. 1148.

FOSTER, P. and YOST, L. (1967) *Buganda rudimentary sedentary agriculture*, University of Maryland, Department of Agricultural Economics, Miscellaneous Publication no. 590.

JOLLY, A. L. (1952) Unit farms, *Tropical Agriculture*, Trinidad, vol. 29, no. 7, p. 12.

JOLLY, A. L. (1955) Peasant experimental farms, *Tropical Agriculture*, Trinidad, vol. 32, no. 4, p. 257.

JOLLY, A. L. (1957) The unit farm as a tool in farm management research, *Journal of Farm Economics*, vol. 39, no. 3, p. 739.

11 Surveys

The only reliable way of collecting information relating to a large number of farms of a particular type is by a survey of a representative sample of them. Many kinds of information regarding a population of farms or farm families or an area of agricultural land may be collected by sample survey. Agricultural sample censuses have already been mentioned as a possible source of farm management data, so too have market price surveys. Food consumption or dietary surveys which involve detailed recording of the quantities of food consumed by a sample of households on specific days may also provide information of some use to the farm management adviser. Land use surveys give valuable information on land resources, their current use and their potential. However, none of these surveys provides all the information needed for farm management analysis and planning. For these purposes a farm management survey is needed. In much of Africa, where subsistence needs are met from farm production, the major source of labour is the family and social constraints are important, both economic and sociological data are needed to describe and explain farming systems. Hence farm management surveys are sometimes called 'socio-economic surveys'.

A socio-economic survey, unlike technical experiments and case-study farms, can provide information in all the four groups listed in the last chapter; namely, description of the pattern of farming, physical input–output data, prices of inputs and outputs and the economic and social constraints. It can therefore provide a complete set of the information needed for analysing farm management problems or planning the reallocation of existing resources. Nevertheless it is desirable to plan the survey in advance; to decide just which items of information are needed and how they will be analysed. Frequently public organizations gather routine information without any clear idea whether it can be used until the point is reached where so much has been collected that it cannot possibly be handled; and yet the collecting process continues. Before embarking on a farm management survey it is desirable, if possible, to carry out a pilot survey of a very small sample to discover which variables are important for inclusion in the main survey and which can safely be omitted.

A sample survey provides not only average or typical data for the type of farm studied but also a measure of the variation between individual farms. For many purposes an estimate of the range of results obtained is more useful than a single average figure. In addition comparisons between the more successful and the less successful farmers may provide important insights into the reasons for success or failure. Methods of analysis of the variation between farms are discussed in the next two chapters.

The two major disadvantages of the survey approach would seem to be: (1) the cost, and (2) that it can only provide information on the existing situation and cannot be used to measure the effects of innovations before they are introduced to farmers. The cost must be much higher than that of a case study if the same information is collected in the same detail on each farm. Where there is a high degree of literacy and particularly where records are already being kept by farmers, it may be possible to conduct the survey simply by collecting record books or by asking farmers to complete a questionnaire. In most of Africa, local people are either not able or not inclined to complete questionnaires, so all information must be collected by enumerators. Suitable people for training as enumerators are often scarce and so, too, are funds to pay them.

In Kenya some farmers were persuaded to keep records for survey purposes, their children filling them in where parents were illiterate. However, we are told that 'all farms chosen for study were atypical, having been drawn from the small group which had shown substantial development from the majority traditional situation'. Certainly if only those farmers who are prepared to keep records are included in the sample this will introduce bias. It was estimated that each set of full business figures collected in Kenya cost, on average, something like £9 per farm, whilst basic agricultural statistics from settlement schemes cost around £5 per farm.[1] Those costs are quite low compared with many other surveys. Thus the survey of Nigerian cocoa farmers cost nearly £70,000 in the early 1950s and although over 350 families were

[1] MacArthur, J. D. (1968) The economic study of African small farms: some Kenya experiences, *Journal of Agricultural Economics*, vol. 19, no. 2, p. 193.

chosen, eventually only 187 families were included in the analysis. The cost per farm family was therefore £100 or more.[2]

Surveys can only provide information on farming systems already in existence. By the time an innovation has been introduced on a sufficient number of farms to provide a statistical sample, it is no longer an innovation. Unit farms are better suited to this purpose.

It is not possible, in a single chapter of a book such as this, to give a complete account of the statistical theory and practical methods of carrying out farm management surveys, but it is proposed to discuss some of the major problems, under the headings— *the average farm, sampling, organization* and *measurement*.

THE AVERAGE FARM

A particular problem with farm management surveys is that of finding a single measure for each characteristic which typifies the whole sample of farms. Naturally, for this to be worth while, the members of the population from which the sample is drawn must be recognizably similar. For instance, it would be meaningless to treat smallholder farms and large commercial estates as if they belonged to the same population. Nevertheless, while there is similarity between the various members of the population, very great differences exist between the largest and the smallest members. Some method of summarizing the varied results is needed.

The simplest and most obvious method is to take the mean value of each characteristic to describe the 'average farm'. The mean, more precisely called the arithmetic mean, is simply calculated by summing the values for the farms in the sample and dividing by the number of farms. This is probably the most common method of analysing survey results but it has certain disadvantages, because the mean is an abstract value; the average farm does not exist in reality.

Firstly, the mean value for discontinuous items such as the number of persons in the labour force or the number of livestock kept is quite likely to include a fraction, which is of course meaningless. For instance, in a village studied in Nigeria, the mean number of wives per farmer was 1·9 and the number of surviving children per farmer was 4·4.[3] The result can be rounded off to the nearest whole number but the rounded value is no longer the true mean.

[2] Galletti, Baldwin and Dina, *Nigerian cocoa farmers.*
[3] Upton, *Agriculture in south-western Nigeria.*

Linked with this is the problem that if 1 farmer in a sample of 30 keeps 20 hens, the mean number of hens per farmer is 0·67. Even if this is rounded up to 1, the idea of a poultry unit consisting of 1 hen may be quite unrealistic.

The second problem, which is really more important, is that there may be a marked lack of symmetry, the farms tending to

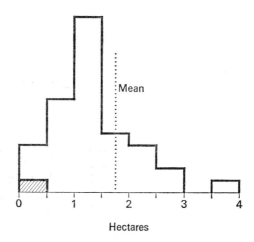

FIGURE 11.1 Frequency distribution of farm sizes. Frequency scale: shaded area represents one farm.

cluster much nearer to one extreme than the other. This gives a skewed distribution like that in Figure 11.1, which shows the way in which farm areas are distributed in the same Nigerian village. The majority of farms are smaller than the mean so the mean is a poor measure of the typical farm area. The same is true for family sizes, for farm incomes, in fact for practically all the characteristics we are concerned with.

Figure 11.2 shows the frequency distribution of numbers of surviving children in the Nigerian village for which the mean value is 4·4. This distribution is again slightly skewed. One way of overcoming this problem and that of meaningless fractions is to use the most commonly occurring value, which is called the mode or modal value. In Figure 11.2 the modal value for the number of children is seen to be 4. This is really a typical value and is realistic in that it is the actual number found on a majority of the farms.

However, the mode is not superior to the mean in all respects. The mode would form a very poor basis for any further calculations of an arithmetical nature, for it has deliberately excluded arithmetical precision in the interests of presenting a typical result. For instance if the mean hours worked by all children per family per year is 746, then the mean hours worked per child per year is 746/4·4 = 169·5. This cannot be estimated from the mode. Furthermore, for a continuous variable, such as land area, there is no single modal value, because every farm in the sample is likely to be slightly different in area from every other farm in the sample. Thus, although there is a precise modal value of 4 in Figure 11.2, the mode in Figure 11.1 is between 1.0 and 1·5 hectares, but cannot be fixed more precisely from the frequency diagram given.

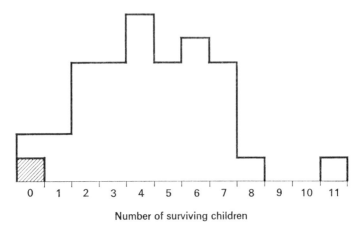

Number of surviving children

FIGURE 11.2 Frequency distribution of family sizes. Frequency scale: shaded area represents one family.

It is always useful to present some measure of the variation between the sample farms. The easiest way is to state the highest and the lowest value obtained; thus the smallest cultivated area was 0·38 hectares and the largest was 3·86 hectares. Alternatively, we might state the range, that is the difference between the highest and the lowest, in this case 3·86 − 0·38 = 3·48 hectares. However, this is entirely dependent on extreme values which may be very unusual freak cases. The mean of the highest 25 per cent of farms, say, and the lowest 25 per cent is less influenced by extreme values.

Only a few of the many measures available have been mentioned but they are sufficient to show that no single measure is satisfactory for all purposes.

An important factor influencing the choice of sampling, organizational and measurement procedures, is the level of precision or accuracy required, when generally speaking increased accuracy involves extra cost. For example accuracy can be improved by increasing the number of farms in the sample but this clearly increases the cost of making the survey. In this connection it should be recognized that there is considerable inherent variation between farms and between years in costs and returns so that a very precise estimate of the average for all farms in the area in one particular year may still prove to be a very poor prediction of the result in another year.

At the same time precision or accuracy is not the only feature to look for in choosing sampling, organizational and measurement procedures. Precision or accuracy is measured by the level of error or uncertainty surrounding a particular estimate, so an imprecise estimate may well turn out to be correct. It is more important that an estimate should be unbiased than that it should be precise because if it is biased then it must inevitably be wrong or misleading.

SAMPLING

The sampling problem is to decide how to select the sample from the population. This sounds, and indeed is, a simple thing to do but unless we ensure that there is no bias involved in our sampling method, there is no hope whatever of our being able to make scientific statements about the population from the knowledge we obtain from the sample. It is by no means easy to ensure that there is no bias.

Suppose, for instance, the agricultural extension service is asked to recommend names of farmers likely to be willing to cooperate in providing farm management data. These farmers are likely to be more progressive than their neighbours and may have introduced new techniques not commonly employed on the majority of farms in the population. If this error is avoided by eliminating these farmers from consideration when selecting the sample, this would be little better, for the bias would be in the opposite direction.

We do not usually know what biases there are in our sampling procedure if we choose it for reasons of mere convenience, speed, or cheapness, or because it has no obvious disadvantages. In sampling it is never enough not to have detected a bias; the sample should be drawn in such a way that no possibility of bias can arise. We are only really safe in this respect if the sample is selected in some way which is completely unrelated to any conceivable variable. To ensure this, we employ a chance mechanism to select the sample, that is we take a random sample. With a simple random sample every farm in the population has an equal chance of being selected.

The random sample is therefore the ideal to be aimed at to avoid bias. However, a random sample is not always possible for farm management surveys. Thus a great deal of information, some of it of a highly personal nature, must be collected over at least one cropping season and preferably longer. This may require many visits by the enumerators and may take up a great deal of the farmer's time. It is therefore essential to find farmers who are able and willing to cooperate. Not all members of a random sample will be agreeable. Furthermore, in many parts of Africa there is no complete list of all the farmers in the population. Without such a list or 'sampling frame' it is impossible to ensure that every farmer has an equal chance of being selected.

For some purposes, such as land use surveys, it is possible to use areas of land (or their equivalent on maps) as the sampling frame, but where, as with a farm management survey, contact with the individual farm families is necessary, the best frame to use is one based on a list of the human population. Such lists may be prepared from the returns of the most recent population census, or, in their absence, from the records of local administrators, tax collectors or a centralized marketing agency. Most of these lists are likely to be either out of date, or incomplete, or both. If no comprehensive and up-to-date information for a sample frame exists, it may be desirable to make a reconnaissance survey of all farms covering only a few items, such as farm area, type of land and family size, in order to compile a complete list of farms in the area. Thus every effort should be made to obtain a complete sample frame and to select a random sample. Where this is not possible, the danger of bias must be borne in mind.

There are possible modifications to the simple random sample in which every farm has an equal chance of selection, although these modifications involve random selection at some point. For the 'stratified random sample', the population is divided into a number of groups or strata. These strata may consist of: (1) administrative units, (2) ecological/agricultural zones, (3) village or farm size groups, or any other means of classifying farms. Within each stratum a random sample of farms is selected, which means that every farm has an equal chance of being selected. This chance, however, might not be equal to that in a different stratum of the population. A stratified random sample is thus, in effect, a collection of simple random samples from a collection of populations.

It is generally the case that a stratified random sample gives more precise results than a simple random sample, especially if the strata are selected so that the variation between strata is as large as possible and hence the variation between farms within each stratum is minimized. The results are more precise, simply because the variation within each stratum is less than the variation in the whole population. However, in order to define the strata it is necessary to have some additional information on the population, besides the sampling frame. This additional information will obviously be available if the sample is to be stratified by administrative units, but this method of defining strata is likely to be less effective in improving precision than stratifying by ecological zone and farm size.

Obviously, since many items are being recorded on each farm, one basis of stratification may not be equally effective in improving precision for each item. For example, the types of crops grown and the area of each crop per farm are likely to differ considerably between climatic zones *but* family sizes or the amount of capital used might vary more between farms within zones than between zones. Unless we are very fortunate, therefore, we must expect the gains from stratification to be relatively modest *but* it will practically always bring about some improvement for every item, no matter what the basis of stratification. There is nothing wrong with using different samples to provide different items of information apart from a possible increase in cost. This method was used in Uboma, in Eastern Nigeria.[4]

The random cluster sample involves dividing the population into a number of groups. A random selection is made from these groups. All the individuals in the chosen groups then constitute a cluster sample. Whereas, with the stratified random sample, all groups or strata are included but only a sample of farms within each group are surveyed, with the random cluster sample only a sample of groups are included but all farms within the sample groups are surveyed. Unless the clusters are very carefully defined so that each one includes as much variation as possible, or reflects the full range of variation in the whole population, this method is

---

[4] See Oluwasanmi and others, *Uboma—a socio-economic and nutritional survey of a rural community in Eastern Nigeria.*

likely to be less precise than simple random sampling for a given sample size. However, its big advantage is that it is likely to be cheaper than other forms of sampling, because the cost of enumerator's travel from one farm to another is much reduced.

Random cluster sampling is particularly useful (1) where there is no population list to serve as a sampling frame, and (2) where there is a large dispersed population or where communications are bad. Cluster sampling was used in a farm survey in Zambia.[5]

Generally speaking, some of the advantages of both techniques can be obtained by means of a multi-stage random sample. For a two-stage sample, the population is divided into a number of groups, villages for example; a simple random selection is made from the groups; then a simple random selection is made from the farms in each selected group. All the individuals selected in this way, taken together, constitute the two-stage sample. Thus the two-stage sample may be viewed as a cluster sample, in which only a sample of the farms within each cluster are studied, or a stratified random sample in which only a sample of the strata are included. Most of the field enquiries in the agricultural sector in developing countries have been based on multi-stage samples. Thus the first stage groupings may be ecological/agricultural zones; the second stage groupings villages; the third stage groupings farms or families; and for some purposes the fourth stage groupings are individual plots.

Where there are no population data available to serve as a sampling frame, ecological zones and villages may be distinguished and sampled from aerial photographs or maps if available. Each village in the sample may then be subjected to a population census in order to provide data for sampling farms at random within the villages.

This very brief review of sampling methods should show that selection is by no means the simple and obvious matter that it at first appears. Before embarking on any survey it is advisable to get the help of a statistician or to study the theory of sampling methods before drawing the sample.[6]

One general point regarding sampling is worth noting, namely that it is *sample size* and *not* the fraction of the population sampled

[5] Bessell, J. E., Roberts, R. A. J. and Vanzetti, N. (1968) *Universities of Nottingham and Zambia Agricultural Labour Productivity Investigation (UNZALPI) Report no. 1: Survey fieldwork*, Nottingham (mimeograph).

[6] A good introductory text is by Stuart, A. (1964) *Basic ideas of scientific sampling*, London, Charles Griffin. More comprehensive works are Yates, F. (1960) *Sampling methods for censuses and surveys*, 3rd ed., London, Charles Griffin; and Cochran, W. G. (1963) *Sampling techniques*, 2nd ed., New York, Wiley.

which almost entirely determines the precision of estimation for a given population. For most purposes a sample size of thirty farms in each independent stratum is probably adequate. There is little point in surveying a sample of a thousand or more farms. Resources would be better used in improving the accuracy of the data collected or in collecting additional data. Even where the number of farms studied is an insignificant fraction of the total population, a random sample of sufficient size can be used to draw reliable unbiased results and to test the accuracy of these results. If, however, it is impossible to draw a random sample then it is important to check as thoroughly as possible whether the results are biased in any way.

Where a survey is made in just a single year or only a few years, the years are in fact a sample from the whole population of an infinite series of years. Random sampling is not possible in this respect so it is important for the investigator to determine to what extent the information gathered each particular year represents normal or average conditions, particularly for crop yields, animal production and price levels. This, of course, does not apply where farm management surveys are made continuously year after year. Indeed, there is much to be said for establishing surveys on a permanent basis. Farm conditions and factors which influence farm businesses are constantly changing. Thus data rapidly become outdated. After a farm management survey has been repeated in the same area for a number of years, the data become more and more accurate, and the time involved and money spent diminish because farmers become more familiar with the nature of the survey and the type of information required. Enumerators become more experienced and do not need to repeat the initial training. Furthermore, data from repeated surveys make it possible to identify trends in yields, prices and factor inputs.

ORGANIZATION

Because farming is seasonal, information for a farm management survey must cover at least one cropping season in order to include a complete sequence of operations. Certain inputs such as labour are used continuously throughout the year, certain outputs such as milk or rubber are produced fairly continuously throughout the year and consumption occurs throughout the year. In order to collect accurate data on these quantities, ideally they should be recorded as they occur. While this may be feasible with a case-study or unit farm, it would clearly be prohibitively expensive to carry out a survey in this way. Hence it is necessary to rely on the farmer's memory of past transactions and productive activities. With daily visits he only needs to recall what happened a few hours earlier, whereas at the other extreme a single-visit survey will require the farmer to recall events from some months earlier.

The single-visit approach has been used in Tanzania by Collinson and Beck for farm management studies, the survey generally taking place just after harvest to minimize the period of recall.[7] Clearly the data on labour use and other items which occur throughout the year are not as accurate as those obtained from more frequent visiting. However, the saving in cost may be sufficient to justify the loss of accuracy, in some circumstances. It may be further argued that the variation between farmers and seasons in such items as rates of working is so great that the observational errors are relatively small and unimportant by comparison. Other studies adopted far more frequent visiting at daily or two-daily intervals.[8] In one of these it is claimed that a senior statistician carried out some accuracy tests for different recording intervals and deduced that there was a marked decline in accuracy after an interval of more than two days. Despite this there is a disadvantage, apart from the cost, of very frequent visiting, in that farmers may become exhausted and bored with the regularity of questioning on the same topics. As an apparently satisfactory compromise, monthly visiting is used in farm business studies in Kenya.[9]

Clearly, the research worker is faced with the problem of resolving the conflict between what is desirable and what is possible with the limited resources available for collecting data. Thus, one of the main problems in designing any survey is the interaction between sample size and visiting frequency. Given a fixed quantity of resources, a rise in the visiting frequency implies a reduction in

[7] Collinson, M. P. (1961) Bukumbi 1960–1; (1962) Part of Usmao chiefdom, Kwimba district, 1961–2; (1963) Luguru ginnery zone, Maswa district, 1962–3; (1964) Lwenge C. S. zone, Geita district; *Farm Management Surveys* nos. 1–4, Tanganyika Ministry of Agriculture, Western Research Centre (mimeographs). Also Beck, R. S. (1963) *An economic study of coffee–banana farms in the central Machame area*, Dar es Salaam, Ministry of Agriculture (mimeograph).

[8] e.g. Heyer, J. (1965) *Seasonal labour inputs in peasant agriculture*, a paper presented to the Second Conference of East African Agricultural Economists, Nairobi; Heyer, J. (1967) *The economics of small-scale farming in lowland Machakos*, Nairobi, University College, Institute for Development Studies, Occasional Paper no. 1; Pudsey, D. (1966) *Pilot survey of twelve farms in Toro Uganda*, Fort Portal, Department of Agriculture (mimeograph); Norman, *An economic study of three villages in Zaria province*.

[9] MacArthur, *The economic study of African small farms*.

the number of farms studied. It may be worth losing accuracy through infrequent visiting if this means that more strata of different farm types can be studied or that a sample large enough to provide statistically valid results can be drawn from each stratum.

For some surveys it has been thought necessary to give incentives to persuade farmers to cooperate. During the Nigerian cocoa farmers survey 'it proved necessary to recompense the cooperating families for their trouble in giving information'. In Northern Nigeria the sample farmers were promised a bag of fertilizer at the end of the survey year. Such gifts or promises should not be necessary provided that farmers are approached through the proper channels, normally through the local council or chief or the extension service, and provided that the purpose of this survey is explained to them. Without such explanation farmers are likely to be suspicious of the aims of the study. In particular they often fear that the government will take over their land for development purposes or that their taxes will be increased. They must therefore be persuaded that there will be no disadvantage to them in becoming involved in the enquiry and indeed that they will be contributing to the development of the whole country. New agricultural policies might follow from the results of the survey and the sample farmers themselves would benefit.

The personality and behaviour of the enumerators has an important effect on the willingness of farmers to cooperate. Confidence must be established between the farmer and the enumerator, and this requires the latter to be tactful and friendly and to respect the social courtesies of the community concerned. It is also highly desirable that he should speak the local language and should know something of local farm conditions and practices, in order to ask questions intelligently and to check, on the spot, the accuracy of the information given by the farmer. For these reasons it may be desirable to recruit enumerators from the area of the survey so that they are already part of the community. There are two difficulties, however. Firstly, the man chosen as enumerator may have made enemies in the past, in his home locality. This would no doubt affect the willingness of farmers to cooperate. Secondly, it is extremely difficult to sack an enumerator who is a member of the community, if he turns out to be unsatisfactory, since he can very easily turn farmers against the scheme, and create difficulties for the *new* man recruited in his place. Ideally, staff should be employed on a permanent basis as they will gradually improve at the work over a period of years. In any case it is desirable that they should be employed for the whole survey period.

The survey may be carried out more cheaply by using part-time enumerators. Village schoolteachers are sometimes used but more commonly field assistants from the government agricultural extension service are employed as part-time enumerators. Such arrangements are rarely satisfactory since their other responsibilities may become more urgent and thereby cause the part-time enumerators to neglect the survey. When field recording is only a supplementary part of their duties, visits tend to become irregular, as the enumerators are not committed to making a success of their recordings. Thus, where possible, enumerators should be employed on a full-time basis. University students may be used during vacations full-time, but they are only suitable for a single-visit survey or one covering only a short period of the year.

Provided that enumerators have the right personality, speak the local language, are employed full-time on the project and receive some initial training and later supervision, educational standards need not be high. Most farm management surveys in Africa have been carried out by enumerators with no more education than seven or eight years' primary schooling. Selection may be based on an interview and perhaps a simple test of ability to write clearly and to make simple calculations. A trial period gives a most useful guide but may be too costly or time consuming to organize.

Opinions vary as to the optimum period of training for enumerators, which must depend to some extent on previous experience. Thus, agricultural extension workers in Kenya embarked on farm management studies after 'no more than a few hours' talk on a single occasion with one of the economists'.[10] In some cases, up to six months' training has been provided. There seems to be some evidence that interviewers and enumerators who get a very short training make no more errors than those who get one of two or three times the length. The most usual period appears to be about two weeks. During this period the enumerators should have the purpose, significance and importance of the survey explained to them. They should study and understand the survey procedure, schedules and questionnaires and they should be taught the techniques of assessing areas and weights and measures and of asking questions. Actual practice in completing schedules and questionnaires on non-sample farms before the survey begins is desirable.

Most of those responsible for directing farm management surveys

[10] MacArthur, *The economic study of African small farms.*

have found that each enumerator can manage to visit two or three farms per day or ten to fifteen per week. However, a recent Zambian survey reports between twenty and thirty visits per day.[11] The number would be less if farmers were widely dispersed and lived in different villages. Thus the total number of farms surveyed per enumerator must depend upon the frequency of visiting. For a single-visit survey spread over say two months, one enumerator might cover over 100 farms, whereas with visiting every two days each enumerator can only manage five or six farms.

Again there are differences of opinion regarding how often the enumerators should be visited by the supervisor. It must depend to some extent on the training and experience of the enumerator. Supervisory visits generally need to be most frequent during the early stages of a survey while enumerators are gaining experience, especially if they have received very little initial training. Furthermore, the research officer responsible for the survey is likely to benefit himself by seeing some of the fieldwork in progress at an early stage and his interest may provide encouragement to the enumerators. The enumerators should make at least two copies of all the information they collect and should send one copy to the survey office as soon as possible. This should then be checked for omissions, obvious errors or inconsistencies, and a query should be sent to the enumerator if any of these are found. The flow of questions back to the enumerators is a great help in keeping them on their toes, provided there is only a short delay involved in office checking.

This implies that there must be sufficient office staff to carry out the preliminary analysis of the survey records as they come in. It would require about one clerical officer on office tabulation to every five enumerators. In many cases too few office staff have been employed so these error checks have not been carried out properly and there has been a long delay before the final results are produced.

Enumerators are usually provided with printed or duplicated forms on which they can fill in the replies to their questions. It is useful to distinguish between two types of form, schedules which simply list or set out, in tables, the items of information required, the enumerator being left to frame his own questions, and questionnaires which consist of a series of questions, to be asked in the precise wording given in the local language. Schedules are suitable for collecting factual information on farm size, family size, labour

[11] Bessell, Roberts and Vanzetti, *UNZALPI Report no. 1*.

use, capital resources, yields obtained and the personal history and circumstances of the farmer. Of course, the enumerator could collect the information on a blank sheet of paper, but the schedule organizes the questioning and ensures that no items are missed. Questionnaires are used in investigating attitudes and aptitudes, when it is desirable that every member of the sample is asked exactly the same questions in precisely the same way. However, survey forms are usually part schedule and part questionnaire.

For both schedules and questionnaires it is important to have some general agreement with enumerators regarding the forms used. For example, formal definitions will be needed for household (usually a group of people who eat together); family (the nuclear family of a man, his wives and children still at home); farm (all land and other resources under the economic control of one farmer); parcel of land (a continuous piece of land forming part of a farm, and surrounded on all sides by land of other farmers, public land or communal land); plot (a continuous piece of land within a parcel, covered by one category of land use such as one crop or one crop mixture)—and so on. Other definitions will be needed depending on local customs and culture. A list should be provided of all the local names for the crops likely to be met during the survey.

Many of the problems of wording and layout of schedules and questionnaires which might crop up may be resolved by pre-testing. This means using the preliminary draft form to interview several farmers from the population to be investigated but who are not included in the survey sample. Pretesting almost invariably leads to improvements in schedules and questionnaires, and furthermore gives an opportunity for practical training of field staff.

MEASUREMENT

It is usually necessary to measure cropped areas of land since farmers generally either do not know the area or, if they do, the local units of measurement are not standard over wide areas. This is important not only to define the quantity of the land resource used but also to define the scale of the enterprise. Thus crop yields and inputs of labour, seed, fertilizer and so on, are usually expressed on a per hectare basis. Grazing land presents a rather different problem in that the land may be actually used communally and the scale of the enterprise may be defined in livestock numbers rather than area covered. Hence the measurement of grazed area is generally not very important and may indeed be impossible if the sample farmer shares the land with others. It is quite likely that water supply or some other factor, not land, will limit the numbers of animals which may be kept and the farmer may be able to state the maximum number he could carry.

Information is required on the area of unused but cultivable bush including bush fallow if the availability of land is likely to limit output. However, this is not often the case in Africa, and where it is, there is unlikely to be much unused land other than temporary bush fallows. The area of bush fallows may be measured in the same way as the area of crops, or it might be obtained more simply, although less accurately, by multiplying the cropped area by the ratio of years of fallow to years of cropping. Thus if the farmer states that he normally crops his land for three years after clearing and then rests it under fallow for six years, the total area of his fallow should be approximately 6/3 times his arable cropped area.

For cropped areas there are frequently problems of locating and identifying plots besides the problems of actual area measurement. Where each farm is made up of many small scattered plots some are easily missed, and where there is a continuous growing season and a sequence of crops during the year, the boundaries of plots may shift from month to month. Hence a sketch map should be prepared for each farm, showing all the plots cultivated by the farmer and the crops grown on each. The map may need modification several times during the year. In order to relate labour and other inputs and outputs to particular plots it is helpful to mark each plot on the farm. Difficulties also arise in defining plot boundaries. The limit of cultivation can be determined at plot preparation time, but once plants are growing this is often exceeded by the vegetative part of the plant; the pumpkin is an extreme example of this effect. Similarly, the canopy area of a tree plot exceeds that of the stems or even the weeded area. These differences, while not very important on large plots, can represent a high proportion of the area of small plots.

If the plots cultivated by each sample farmer are identified on the ground, large scale (low level) aerial photographs may be used for estimating crop areas, but the costs are very high and it would be difficult to justify repeating the photography for each cropping season.[12] Measurement on the ground involves either dividing each

[12] e.g. Norman, *An economic study of three villages in Zaria province.*

plot up into triangles or parallel-sided strips, or using a compass to take bearings along the sides of the plot.[13] Triangulation is probably the simplest and cheapest method, and if the land is divided up into different sets of triangles as in Figure 11.3 cross checking is possible. Irregularities in plot shape are ironed out by a give and take process, leaving as near as possible an equal area of crop outside the measuring lines as there is bush inside the lines.

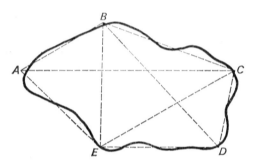

FIGURE 11.3 Area measurement by triangulation. The sum of the areas of triangles *ABC*, *ACE* and *CDE* may be compared with the sum of the areas of triangles *ABE*, *BDE* and *BCD* for checking purposes.

The lengths of the sides of each triangle may be measured by pacing, measuring chains, tapes or knotted cords, or by means of a measuring wheel. For chains, tapes or knotted cords two men are essential, whereas pacing and use of the wheel are possible with only one man although others can help in marking out the corners of the plot. When pacing or using knotted cords it pays to check the length, of, say, 100 paces, or the cord, against a chain or a tape periodically.

When the measurements have been made the areas can be calculated directly from the lengths of the sides of each triangle, using the following formula:

$$\sqrt{s\,(s-a)\,(s-b)\,(s-c)}$$

where *a*, *b* and *c* are the lengths of the sides of the triangle and

$$s = \frac{a+b+c}{2}.$$

[13] Methods of area measurement are discussed in Stewart, F. R. and Grassie, J. C. (1947) *Surveying for agricultural students and planters*, London, George Allen and Unwin; Upton and Anthonio, *Farming as a business*, appendix 2; and Debenham, F. (1961) *Mapmaking*, 3rd ed., Glasgow, Blackie.

Alternatively, each plot may be drawn accurately to scale and its area estimated by one of several available methods. For instance, it may be divided up into squares each representing say one square metre also drawn to scale. The area is estimated by counting the squares.

There is a certain degree of skill involved in measuring plots by any of the available methods so there is something to be said for making one of the enumerators a permanent plot measurer, which will give him an opportunity to develop the particular skills needed.

Mixed cropping raises special problems of measurement to which no entirely satisfactory solution has been found. Where, as is usually the case in Africa, crops are interplanted in mixtures, it becomes very difficult, if not impossible, to estimate the resource inputs used on each individual crop in the mixture. Thus the problem is not only that of assessing the area of land devoted to each, but also that of allocating labour and other inputs used in clearing, cultivating and weeding. One possible approach is to treat each component crop in the mixture as though it occupied the whole area. This implies that one hectare of maize, cotton and groundnuts mixed will yield the same quantities of each crop as would three separate hectares, one of maize, one of cotton and one of groundnuts, in pure stand. Clearly this makes no allowance for competition between the crops, or for variation in their proportions in the mixture. The total area of crops, if the statistics are handled in this way, would naturally be treble the actual area of land under cultivation.

Alternatively, an attempt might be made to divide the area under mixed crop between the individual crops concerned on some concept of the 'proportion of the area occupied' by the respective crops. This requires an assessment of relative ground cover or plant densities of the various crops in the mixture. It may be possible, with practice, to judge the proportionate coverage of the ground, but where there are major differences in growth habit and height of the different crop plants this will be difficult. Relative plant densities generally need to be compared with plant densities in the pure stand of each crop, in order to reflect their relative importance in the mixture, as the normal spacing must vary from one crop to another. The ratio of the recorded density to the plant population density of a pure stand is then used to estimate the proportion of the total area devoted to each crop. To take plant counts requires sub-sampling of the total crop area, as may assessments of relative ground cover. This adds to the enumerators' problems. Furthermore, any assessment of the relative importance of the crops in the mixture is somewhat dubious and ignores the interactions between the different crops in a mixture.

Perhaps the simplest and most acceptable method on theoretical grounds is to treat each crop mixture as a particular enterprise or activity (see chapter on linear programming). It may be possible to describe each mixture in terms of principal crops and secondary crops or it may be possible merely to refer to them as including the several crops. This approach is the most satisfactory where there is some standardization of crop mixtures but breaks down where, as is commonly the case, there is an infinite variety of different crop mixtures, varying from one plot to the next and from one farmer to the next. Alternatively it may be possible to estimate the area of individual crops from information on the quantity of seed used.

The need for frequent visiting, to collect accurate labour records, has already been mentioned. Ideally these records should provide information on the daily labour inputs on each crop and livestock enterprise. It has generally been found that the best way of doing this is to record what each member of the family and each hired worker has done, on which plot or enterprise and for how many hours. In order to get a more complete and accurate picture it is helpful to record all activities both on and off the farm. Only when the farm data are available as part of a complete record of the use of time will a reasonably true figure for labour input on the farm he obtained. This is, in part, due to the very blurred distinction between farm and non-farm activities; for instance, is threshing stored millet a farm activity or part of cooking? Post-harvest operations, such as shelling groundnuts or drying coffee berries, cause difficulties as they may be performed while the wife is cooking and is therefore not available for work in the fields. Records of non-farm activities are also necessary for any assessment of labour availability as opposed to labour use. Having said this it must be recognized that farmers may object to this degree of cross-questioning. One solution is to collect detailed labour records from a very small sub-sample of farms which is changed from week to week. Provided each sub-sample is selected at random, the fact that records are collected from different farms in different weeks should not raise problems.

Measurement of capital inputs requires that an inventory or list of all capital resources available to the farmer should be made at the beginning of the recording period. As a check it is advisable to make another inventory at the end of the period. The inventory would include details of permanent crops, livestock, buildings, tools, equipment, machinery, stores of food, seed and so on and, if possible, cash reserves. For each of these items it is desirable to record not only the physical quantity (e.g. area of trees, number of livestock) but also the age and physical condition, since these factors must influence both the productivity and the value of each asset. In addition the use of capital resources should be recorded through the year, in particular the use of machinery, stores of food, seed and other materials and cash reserves. For permanent crops and livestock, production and disposal should be recorded. Where permanent crops are grown entirely for sale, cocoa for example, a record of sales is adequate, but where home consumption is involved, as with bananas, oil palm, kola, coconuts, and so on, it is more difficult to estimate total production. It may be possible, in some cases such as bananas or oil palms, to count the number of bunches on each tree, or on a sample of trees before they are harvested, and then to weigh a sample of these bunches to arrive at an average weight per bunch. Livestock production and disposal is estimated from records of births, purchases, gifts, losses, slaughterings for religious rites and for consumption and sales. Eggs and milk are special cases which require particularly regular and detailed recording.

Harvesting practices for arable food crops give rise to problems in both labour recording and yield estimation. Food is often taken back to the homestead after work is finished as and when required, so the total harvest may be spread over many weeks or even months. Rather than attempting to measure crop yields by recording and totalling the many small amounts harvested, the use of crop cutting techniques has been advocated. For this purpose the enumerator harvests and weighs a sample of small measured areas from each of the major crop mixtures grown on the farm. Clearly, this is very time consuming and the farmer may not agree to having his crop cut in any case. Furthermore, questions arise as to the stage of maturity at which the crop should be cut and the level of inaccuracy incurred by the measurement of total yield as opposed to the harvested yield which results from farmers deciding after a certain point that the rest of the crop is not worth harvesting.

The other method of yield recording, which consists of totalling the quantities *used* in different ways, gives rise to all sorts of complications in relating the yields to specific plots and in combining products at different stages of maturity and processed to different levels. Whatever method of yield estimation is used, it is desirable to record all sales and purchases to provide a cash balance and prices for evaluating inputs and outputs.

Finally, it is desirable to try to measure social and cultural constraints, to assess farmers' aims and objectives, their attitude to change in general and to specific innovations. Generally some light can be thrown upon these aspects by careful questioning. Perhaps a sociologist is better qualified than a farm management economist to investigate these matters *but* where no sociologist is available the farm management researcher is justified in raising such questions. It is often useful to ask about a particular attitude in several different ways in several questions, rather than in a single direct question. This enables the researcher to build up a clearer picture of each attitude and sometimes to assess the relative strength with which the attitude is held. In any form of questioning it is clearly important to avoid suggesting answers to the informant. Questions which do this, such as 'Do you grow your own food crops in order to avoid the risk of food shortages?' are known as leading questions.

Ideally, to avoid influencing the answers given, open ended questions should be used. That is, the farmer should be left free to give any answer that comes into his head. However, there is always the danger that fifty different farmers will give fifty different answers which makes it difficult to summarize the result. Analysis is easier if the farmers are offered a limited choice of replies, such as 'I grow food crops because (1) it is profitable to do so, (2) it avoids risk or (3) it is customary to do so.'

This, in a sense, is now a leading question. There is a great danger that some important alternative answer has been missed in the preparation of the questionnaire, or that the farmer will be influenced by the choice of answers offered.

SUGGESTIONS FOR FURTHER READING

CLEAVE, J. (1965) *The collection, analysis and use of farm management data in Uganda*, East African Agricultural Economics Society Conference paper (mimeograph).
HUNT, K. E. (1969) *Agricultural statistics for developing countries*, Oxford, Institute of Agrarian Affairs.
YANG, W. Y. (1965) *Methods of farm management investigations*, revised ed., Rome, F.A.O., Development Paper no. 80.
ZARCOVICH, S. S. (1965) *Sampling methods and censuses*, Rome, F.A.O.

12 Farm business analysis

PRESENTATION OF RESULTS

Once farm management data have been collected they must be processed into a form suitable for use by development planners, technical research workers, extension officers and others. Data obtained from surveys or case studies must be analysed or broken down into individual items, then summarized into a usable form. When data are obtained from technical experiments or secondary sources such as published material or local experience the whole farm picture must be built up or synthetized. Once the data have been processed they should be published as soon as possible since the information rapidly becomes out of date. Prompt publication not only increases the usefulness of the results and increases the likelihood that they will be used, but also creates goodwill among those who have financed the study and those who have contributed their time and effort to it.

The important numerical results of any farm management investigation will be the quantities of productive resources available, the inputs used and the outputs obtained. For purposes of analysis and planning, information is needed on the relationship between the inputs used and the outputs obtained. It is sometimes possible to estimate the production function, relating inputs and outputs from experimental or survey data (see the next chapter), but where this is not possible or inappropriate then it is necessary to relate inputs and outputs to some common unit for measuring size of enterprise. For crops this is usually the hectare, so that yields are expressed per hectare as are labour and other inputs. This is probably the most convenient way of relating inputs and outputs but it does have a possible disadvantage in placing undue emphasis on costs and returns per hectare. Where there is surplus land, return per hectare is a poor guide to relative profitability of different crops. Inputs and outputs for tree crops are sometimes expressed per tree, though this has the disadvantage that the quantities involved may be very small. Livestock inputs and outputs are usually expressed on a per head basis. There is an added advantage in expressing inputs and outputs on a per hectare, per tree or per head of livestock basis in that it enables different farms with enterprises of different sizes to be compared. It should be noted, however, that this approach tells us nothing about marginal products; it simply provides estimates of overall average product.

The following is a list of suggestions of the type of physical data it is desirable to obtain from a farm management investigation. It is not always possible to provide all the information listed, while some studies have given data not included here, but for most purposes this list is appropriate.

(1) *Land.* Farm area, plot sizes and distances from the home, tenure and method of acquiring land, area of individual crops or mixtures, rotations and other fertility maintenance practices.

(2) *Labour.* Family size, total hours worked, hired labour use, distribution of labour used by seasons and by enterprises, division of labour between family members and hired labour, sex/age specialization, social commitments, rates of working.

(3) *Capital.* Stocks and reserves of food, cash etc., livestock numbers, ages and condition, tree crop area or numbers, ages and condition, buildings, tools, machinery and equipment.

(4) *Production.* Total yield obtained, quantity sold, quantity consumed as food, quantity used as seed or livestock feed.

INCOME DATA

In order to compare returns from different enterprises or to deduct costs from output we need some common unit of value. For most purposes market prices are the most useful and convenient, except in pure subsistence farming where none of the product is sold (see Chapter 1). Records of sales of crops and livestock may be used to arrive at an estimate of total cash income. If the cost of purchased farm inputs is deducted, this gives the disposable cash income which may be used for consumption or investment or else saved. This is not a good measure for comparing incomes, since home consumed food is also part of the family income. Hence, in order to estimate total income and to compare the returns from different enterprises we must consider the total production or 'gross output'.

As we have seen, totalling the quantities used in different ways may not be the best method of arriving at gross output. It is generally easier to apply the current market price to the total recorded yield of each crop and livestock enterprise, though there is the disadvantage that this ignores wastage and storage losses. Whichever method of estimation is used, gross output frequently needs adjustment for changes in the amount of produce on hand at the beginning and end of the accounting period. Thus produce left over from the previous year should not be counted as part of the current year's output, while produce left over at the end of the current year should be included. If any produce is purchased this should not be included in gross output. Hence it is necessary to deduct purchases of produce from the total quantity used. The calculation of gross output may be illustrated with a few simple examples as shown in Table 12.1. Thus, in summary we can say that

TABLE 12.1 Enterprise gross outputs

	Product disposal			Product inputs		Difference equals gross output
Enterprise	On hand at end of year	Consumed	Sold	On hand at start of year	Bought	
Millet (bags)	3	10	1	2	4	8
Cassava (tons)	1	7	—	2	—	6
Cattle (£)	150	—	30	120	10	50

gross output equals quantity harvested or used during the year plus quantity remaining at the end of year minus quantity bought during the year minus quantity available at beginning of the year, for each enterprise.

Purchases of crop produce or livestock are therefore included in the calculation of gross output but there may be other explicit costs to consider such as seeds, fertilizers, sprays, hired labour, tools, machinery and fuel, livestock feeds and veterinary bills. Of these seeds and livestock feeds are themselves farm products and may not be recorded separately from other crop purchases. However, where they are recorded separately they may be deducted from the whole farm gross output to arrive at the farm 'net output'. Where no seeds or feeds are purchased, net output is equal to gross output. For some purposes net output is a more useful measure of farm production than gross output, because purchased seeds and feeds are agricultural products which have been brought into the farm from elsewhere.

If other cash expenses are deducted from net output we are left with an estimate of 'social income' which is the return to all the factors of production provided directly by the family. This family income should therefore cover the implicit costs of family land, labour, capital and management. In some studies attempts are made to allocate the farm income between these family resources or to estimate the implicit costs of each. However, this can only be done on some fairly arbitrary basis.

In making these calculations it should be remembered that purchases or sales of fixed capital assets should not be included in costs. Instead some estimate of annual depreciation should be used.

ALLOCATION OF COSTS

A farm business is an organic unit, and it should not be considered as separate departments or isolated segments of individual crops or livestock. Yet there is often a need to make an analytical study of the constituent parts of the farm business in order to understand the internal structure of the various enterprises and their relationships to the farm organization as a whole. Since total farm gross output is calculated by adding together the gross outputs of the individual enterprises there is no problem in estimating the relative contribution of each enterprise to the total. Where a crop mixture is being treated as a single enterprise, naturally the gross output is the sum of the outputs of the constituent crops.

Problems arise however in deciding how to allocate costs between enterprises. Clearly the cost of maize seed should be charged against the maize enterprise, cocoa sprays should be charged against the cocoa crop and poultry feed against the poultry enterprise. Other costs, such as that of family and regular labour, are not so easily allocated because they are fixed or indivisible. Even where detailed records are kept, showing the number of man-hours spent on each enterprise, it is unrealistic to charge so many man-hours at a fixed hourly rate to each enterprise. This ignores the possibility of complementary and supplementary enterprises. The fixed labour requirement is determined by the system as a whole, not by the size of an individual enterprise. The same applies to the fixed cost of an ox team or a tractor.

Hence the allocation of fixed costs to individual enterprises is not only unnecessarily time consuming but also illogical. Only the variable costs need be allocated to individual enterprises and deducted from the enterprise gross outputs to give what are usually called 'gross margins'. The total gross margin from all the enterprises minus the total fixed cost then gives a measure of farm profit. The logic behind this approach is that the fixed costs are more or less independent of the amount produced, and the important thing is to be able to calculate the contribution of each enterprise towards covering these fixed costs and leaving a margin of profit. It should be noted that the term 'margin' used in this sense simply means a difference between costs and returns. In fact the gross margin is sometimes called the gross profit.

There is no hard and fast distinction between fixed and variable costs. As pointed out earlier, the area of tree crops on a farm is best considered as fixed for a single season, but for long term plans is clearly variable. Thus other distinctions are made between the costs which should be deducted from gross output in calculating the gross margin and those which should not. All the following distinctions are sometimes made though they do not all lead to the same classification:

Variable costs—fixed costs.
Avoidable costs—unavoidable costs.
Stock resources—flow resources.
Specific costs—joint costs.
Direct costs—common costs.
Prime costs—overhead costs.

Variable costs are usually also avoidable. For instance the cost of seed and fertilizer used for maize production varies with the quantity of maize planted. The cost can be avoided entirely if *no* maize is produced. On the other hand fixed costs once incurred cannot be avoided so easily. Once an extra regular worker has joined the family labour force, the cost of employing him has been incurred no matter how productive he is. His cost cannot be avoided by stopping production of maize, for example.

The difference may be made clearer by distinguishing between stock and flow resources. Resources which are available in the form of stocks, such as seeds, fertilizers and sprays, can be stored. If they are not used at a particular point in time they can be kept for future use. Generally these stocks can be divided up into small units of fractions of a bag or a tin. Because they can be divided up into small units they are variable, and because they can be stored for future use, the cost to a particular enterprise is avoidable. Resources such as regular labour or machinery, on the other hand, provide a continuous flow of man-hours or machine-hours which cannot be stored up for future use in the way that seeds can. Unused labour in January will not add to the labour supply at harvest time in August. The cost of the flow is fixed and unavoidable whether the labour is actually used at a particular time of year or not.

Alternatively we may distinguish between specific costs or direct costs which are clearly attributable to a particular enterprise and joint or common costs which apply to the farm as a whole. In many cases specific costs are variable and joint costs are fixed, but this is not always so. Specialized machines and equipment developed for a particular crop (a groundnut huller for instance) are specific to that crop but they are also fixed, since they do not vary directly with the quantity of crop grown. Such specific, fixed costs would not normally be deducted from the gross output in calculating the gross margin. The distinction between prime costs and overheads roughly corresponds with that between specific and joint costs, although it is based on accounting convenience. Overhead costs are those which it is inconvenient to allocate to individual enterprises.

The usual classification of costs into the variable and fixed categories is as follows:

VARIABLE COSTS	FIXED COSTS
Crops	Rent if any
Seeds	Wages if any
Fertilizers	Interest on loans if any
Sprays	Depreciation of machinery, equipment and buildings
Livestock	Maintenance and repairs of same
Livestock feeds	
Veterinary costs	

Some costs raise special problems of classification. Depreciation of trees and livestock are specific costs which are relatively fixed in the short run. However for most purposes and particularly for farm planning these costs are best treated as variable costs. Tractor fuel costs vary directly with the amount of tractor use but in order to avoid a lot of extra recording it is common practice to put them with the fixed machinery costs in one inclusive figure. Temporary hired labour probably raises the biggest difficulties of all. Strictly speaking the cost is variable, avoidable and specific to a particular enterprise, as for cotton picking for example. However, if different farms are to be compared, or averaged, and some use temporary labour and others don't, problems arise. Those farms where temporary labour costs are deducted will have lower gross margins than those where the labour force is fixed. The treatment of temporary hired labour costs therefore depends upon the circumstances. Where labour is rarely hired on a temporary basis and then only on a few farms it is probably best treated as a fixed cost. However, where it is common to hire temporary labourers for a specific task such as cotton picking, then the cost should be treated as a variable cost, specific to a particular enterprise.

If fixed costs do not alter much with changes in production then where total gross margin can be increased, farm profit will rise. If the increase in gross margin can be achieved with the existing supply of fixed resources and hence the existing level of fixed costs, the farm profit will be raised by exactly the same amount as the gross margin. For this reason it is possible to plan changes in the farm system in terms of gross margins alone and leave fixed costs out of the calculation. In fact in many parts of Africa, the family farmer does not incur explicit fixed costs. He pays no rent, nor wages to his family who make up his regular labour force, he has hardly any buildings and equipment and does not borrow much capital. Practically all the African farmers' costs are variable. This means that practically the whole of the total gross margin represents family or social income.

COMPARATIVE ANALYSIS

The data on physical inputs and outputs and cash gross margins discussed above, together with information on technical and social constraints, are sufficient for most planning methods. However, they apply to the typical, average or modal farm. In practice results vary from farm to farm and it might be interesting and useful to analyse the causes of this variation. For instance a low total gross margin on a particular farm may be due to an unduly low ratio of land under high-value crops, to little mixed cropping or too much fallow land, to low yields, low prices or high variable costs, and it would be useful to distinguish which of these factors is the main cause. However, results are only low or high in relation to those obtained on other farms and not in any absolute sense. Thus the analysis of the causes of variation in results depends upon comparisons between different farms. The approach is therefore known as comparative analysis. Unlike forward planning, it is, of course, concerned with past results, but it is hoped that the analysis of past results may give useful guidelines for the future.

For several years, comparative analysis was used by the agricultural extension services in the United Kingdom as the basis of their farm management advice. The records and accounts for individual farms receiving advice were analysed to give so-called 'efficiency factors' which were then compared with standards which were usually the average results obtained from a sample of similar farms. The procedure was intended to show up weaknesses in the management of the individual farm which then served to guide the farmer as to where he should concentrate his attention in the future. More recently, in the United Kingdom, some farm economists have used planning techniques, rather than comparative analysis, to suggest improvements on the individual farm. In any case both techniques are clearly unsuitable for much of Africa, depending as they do on individual advice to farmers who keep detailed records and accounts.

However, some useful additional information may be obtained from surveys by comparing different characteristics of the high income and low income farms in the sample. This technique was used to analyse the results of a survey in two villages in the derived guinea savanna zone of Nigeria.[1] The average annual income on twenty-two farms was £46 but the variation between farms was very wide. Family income ranged from £6 to £195 and income per man from £2 to £195. Several of the poorer farmers had a secondary occupation which supplemented the income from farming.

The farms varied in total area ranging from 3·28 to 9·10 hectares but a comparison of family incomes on large and small farms showed no significant relationship between farm area and farm income. Since all twenty-two farmers worked under similar conditions on closely similar farms and income does not appear to vary with area, it would seem that most of the variation in farm incomes must be due to differences in the way the farms were managed.

Some variation in incomes was possibly due to differences in the area of cocoa owned. In fact, only eleven of the farmers had cocoa plots and for these eleven, cocoa provided 72 per cent of total net

[1] See Petu, D. A. and Upton, M. (1964) An economic study of farming in two villages in Ilorin emirate, *Bulletin of Rural Economics and Sociology*, Ibadan, vol. 1, no. 1, p. 1; Upton, M. and Petu, D. A. (1966) A study of farming in two villages in the middle belt of Nigeria, *Tropical Agriculture*, Trinidad, vol. 43, no. 3, p. 179; also Upton, M. (1964) A development of gross margin analysis, *Journal of Agricultural Economics*, vol. 16, no. 1, p. 111.

farm income. These farmers had higher incomes than those without cocoa plots, the averages being £60 and £33 respectively. However, this did not explain much of the variation in net farm incomes. The total gross margin from arable crops (not tree crops) was then calculated and this ranged from £2 to £58 in total, from £0·75 to £11·90 per hectare, and from 8p to 38p per man-day. Further analysis was concerned with discovering the causes of the variation in gross margin per hectare and gross margin per man-day.

The first step in this analysis was aimed at separating the effects of the farm system or choice of enterprises and the efficiency of production of the chosen enterprises. For this purpose it was necessary to calculate a potential total gross margin for each farm. This was done by multiplying the number of hectares of each crop grown on the farm by the standard gross margin per hectare of that crop. These standard gross margins were simply the average for all farms in the sample. The standards used and the estimate of the potential total gross margin for a single farm are shown in Table 12.2.

Thus the potential gross margin per hectare on this farm was £33/4·49 = £7·35 which may be compared with the overall average gross margin per hectare for the whole sample of £5·80. Potentially

TABLE 12.2 Calculation of potential total gross margin for example farm

		£		
	I	II	III (I × II)	IV
Crop	Hectares	Standard gross margin per hectare	Total	Actual gross margin per hectare on example farm
Yams	0·48	46·50	22·32	35·15
Maize	0·95	6·65	6·32	7·30
Cowpeas	0·38	5·70	2·17	5·20
Sorghum	0·02	3·70	0·07	2·95
Cassava	0·05	14·60	0·73	18·60
Cotton	0·04	14·35	0·57	13·60
Vegetables	0·04	20·50	0·82	20·80
Total including fallow	4·49		33·00	
Total gross margin actually achieved			28·14	

the gross margin per hectare for this farm is considerably higher than average. An index of potential is calculated by expressing the individual farm figure as a percentage of the group average thus: ($£7·35/5·80$) × 100 = 127 per cent. The actual performance, that is the total gross margin obtained, is given at the foot of Table 12.2. This figure of £28·14 is considerably lower than the potential figure of £33, which means the performance is below average. Again an index may be calculated by taking the actual total gross margin as a percentage of the potential total gross margin thus: ($£28·14/33$) × 100 = 85 per cent. Similar calculations were made for each of the sample farms so that the average index of potential and performance for high gross margin farms could be compared with the average index of potential and performance for low gross margin farms.

The analysis was taken further by investigating the factors which influence the potential gross margin per hectare and the performance. The potential gross margin per hectare depends upon the percentage of land under fallow, the extent of mixed cropping and the percentage of high gross margin crops, in this example yams, groundnuts and vegetables.[2] Performance is influenced by yields, prices and variable costs.

The results of a comparison of the eleven farms with the highest gross margin per hectare with the eleven farms with lower gross margins per hectare are given in Table 12.3.

The difference in index of potential between the high gross margin per hectare farms and the low gross margin per hectare farms is much larger than the difference in index of performance. This analysis suggests that the index of potential, that is the farming system, is most important in determining the gross margin per hectare. Of the three components of the index of potential, the percentage of mixed cropping seems to have the largest effect.

A similar comparison was made between farms with a high gross margin per man-day of total labour input and farms with a low gross margin per man-day. Actual gross margin per man-day depends upon (1) potential gross margin per man day, (2) crop and livestock production performance and (3) labour efficiency defined as the total standard man-day requirement as a percentage of the actual man-days worked. The results of this comparison given in Table 12.4 suggest that the farm system has little or no effect on

---

[2] 'Mixed cropping' means both intercropping and sequential cropping in this context. Percentage mixed cropping is calculated as the sum of the area equivalents of individual crops as a percentage of the actual cropped area.

TABLE 12.3 Comparison of farms with high and low gross margins per hectare

| | Means of farms with: | | |
Item	High gross margins per hectare	Low gross margins per hectare	Significance or difference*
Area in hectares	3·84	3·69	Not significant
Index of potential	134·0	65·0	Very highly significant
Percentage fallow	74·4	79·5	Significant
Percentage mixed cropping	191·0	113·0	Highly significant
Percentage intensive cropping	26·3	31·1	Not significant
Index of performance	106·0	93·0	Not significant
Index of yield	102·5	101·3	Not significant
Index of prices	97·5	95·6	Not significant
Index of costs	90·3	107·3	Not significant

TABLE 12.4 Comparison of farms with high and low gross margins per man-day

| | Means of farms with: | | |
Item	High gross margins per man-day	Low gross margins per man-day	Significance or difference*
Area in hectares	4·02	3·53	Not significant
Potential margin per man-day	4·4	4·4	Not significant
Percentage fallow	77·0	77·0	Not significant
Percentage mixed cropping	179·0	125·0	Not significant
Percentage intensive cropping	26·0	31·5	Not significant
Index of performance	117·0	81·0	Very highly significant
Index of yield	107·5	96·3	Significant
Index of prices	97·5	95·5	Not significant
Index of costs	81·7	116·1	Highly significant
Labour efficiency	107·0	85·0	Significant

* These are the results of tests of the statistical significance of the difference between the two means. 'Not significant' implies that the difference is small in relation to the remaining variation within the groups.

gross margin per man-day. Performance in terms of outputs, costs and labour efficiency achieved account for the differences in gross margin per man-day.[3]

Many other indices or efficiency factors have been used and no doubt others could be thought up. Frequently they take the form of a ratio between inputs and outputs such as fertilizer purchases per £100 crop output, area cropped per man equivalent or live-stock units per hectare of grazing land.

It is possible to carry the analysis further and attempt to assess the influence of different production techniques on the gross margin per hectare of an individual crop. This was done in a study of progressive farming in central Malawi where gross margins per hectare of tobacco, maize and groundnuts were compared on farms using different cultivation practices. As an example, results for tobacco are shown in Table 12.5.

TABLE 12.5 Factors influencing gross margins from tobacco

Timing of planting	Application of fertilizer and/or farmyard manure	Number of farmers	Average gross margin per hectare (£)
Early	Fertilizer plus FYM	30	50·28
	Fertilizer only	9	42·52
	FYM only	2	55·04
	Neither	1	39·53
Late	Fertilizer plus FYM	13	25·05
	Fertilizer only	12	10·18
	Neither	5	6·67

Source: Hoffmann, H. K. F. (1967) *Case studies of progressive farming in central Malawi*, Malawi Government, Blantyre.

DISADVANTAGES OF COMPARATIVE ANALYSIS

The disadvantages of comparative analysis stem from the fact that it is concerned with average relationships rather than marginal ones. In fact in most comparative studies, gross output per hectare is used as the measure of success. The analysis is aimed at finding

---

[3] Similar results were obtained in a study of three villages in Zaria province, Nigeria: see Norman, *An economic study of three villages in Zaria province: Part 2, Input–output relationships*.

causes of variation in output per hectare and relating costs, both fixed and variable, to the output achieved. Clearly, although agricultural scientists may use gross output per hectare as a measure of technical efficiency, it is a very poor measure of economic efficiency.

In the first place a high gross output may be associated with low profits if costs are also high. Thus gross margins give a better measure of relative profitability of different enterprises than do gross outputs, since variable costs are deducted in calculating gross margins. In the examples quoted above gross *margin* per hectare was used in preference to gross *output* per hectare as a measure of success. However, no matter which of these two measures is used, a second question arises, namely whether the comparisons should be made on a *per hectare* basis. It is clearly desirable, when comparing farms and enterprises of different sizes, to reduce them to some common unit, but use of the hectare as this unit may place undue emphasis on the returns to land as a factor of production. Where there is surplus land available for cultivation, return per hectare is quite unimportant, and return to capital or available labour would be a better measure of economic efficiency.

Unfortunately the measurement of return to capital or available labour presents problems which do not arise in the case of land. Because there is a delay between investment in a capital asset and the return on the investment and because the value of an asset varies over its productive life it is not satisfactory to express return on capital as a simple ratio. Ideally discounting procedures should be used to assess efficiency in the use of capital. In the case of labour there is the problem of seasonality of requirements. Thus gross margin per man-day of total labour use tells us nothing of how the labour use was distributed over the year. Gross margin per man-day of labour in a particular peak month is a better measure of efficiency of labour use. Then there may be several critical labour peaks. The real problem is that production is rarely limited by one single resource constraint operating alone. Where there is more than one resource constraint, the use of a single efficiency index relating to only one of the resources can be positively misleading.

To illustrate the problem let us assume just two resource constraints, land and total labour, though this is no doubt an oversimplification. Now in order to increase output or margin per hectare of land, labour inputs per hectare should be increased. However, in order to increase output or margin per man-day of labour, less of this resource should be used per hectare. Thus we have the technologist's dilemma: that the use of average products as measures of efficiency can lead to diametrically opposed policies depending on the choice of the measure.

This is illustrated in Figures 12.1 and 12.2 for the two resources of land and labour. Figure 12.1 is a hypothetical response curve for increasing labour use on a fixed area of land. Figure 12.2 is a

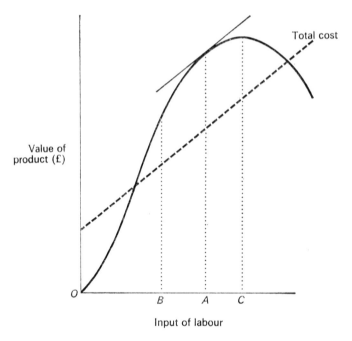

FIGURE 12.1 Choice of technical optimum. Point *A* is the economic optimum (profit is maximized), point *B* is the technical optimum for labour use (average product per man-day is maximized) and point *C* is the technical optimum for land use (average product per hectare is maximized).

hypothetical isoquant showing the combinations of land and labour which will yield a given total product. Similar relationships would be obtained whether product is measured by gross output or gross margin. Now if the objective is to maximize profits there is a single optimum level of labour use at point *A*. Let us assume that a farmer is operating at or near this level. However, if the objective is to maximize land efficiency or average product per hectare, he should increase labour use up to point *C*. If, on the other hand, the objective is to maximize labour efficiency or average product per man-day, he should reduce labour use down to point B.

For a farmer using less than the economic optimum amount of labour, any attempt to increase the average product per man-day by saving labour will in fact reduce profit. Likewise if a farmer is using more than the economic optimum amount of labour per

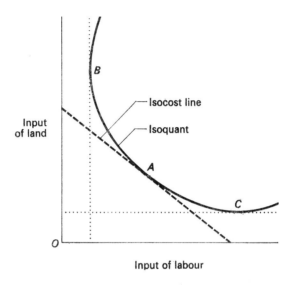

FIGURE 12.2 Factor proportions and productivity. Labour productivity is maximized at point B whereas average product per hectare is maximized at point C, for a given level of output.

hectare, the use of average product per hectare as a measure of efficiency, by encouraging increased labour use per hectare, is likely to reduce profit. In this way the use of measures of technical efficiency such as average product per hectare or per man-day can be misleading.[4]

In much of Africa the use of output or margin per hectare as a measure of efficiency by agricultural scientists may have had a positively harmful effect in encouraging intensive methods of production where extensive methods would be appropriate.

[4] For further discussion see Candler, W. and Sargent, D. (1962) Farm standards and the theory of production economics, *Journal of Agricultural Economics*, vol. 15, no. 2, p. 282.

SUGGESTIONS FOR FURTHER READING

BLAGBURN, C. H. (1961) *Farm planning and management*, London, Longmans.

CANDLER, W. and SARGENT, D. (1962) Farm standards and the theory of production economics, *Journal of Agricultural Economics*, vol. 15, no. 2, p. 282.

GILES, A. K. (1962) *Gross margins and the future of account analysis*, University of Reading, Department of Agricultural Economics, Miscellaneous Studies no. 23.

NORMAN, L. and COOTE, R. B. (1971) *The farm business*, London, Longmans.

OGUNFOWORA, O. and OLAYIDE, S. O. (1969) Assessment of economic efficiency in flue cured tobacco production—a case study of the River Valley Estate mixed farming project, *Nigerian Agricultural Journal*, vol. 6, no. 1, p. 3.

UPTON, M. (1964) A development of gross margin analysis, *Journal of Agricultural Economics*, vol. 16, no. 1, p. 111.

13 Production function analysis

The production standards discussed in the last chapter are useful for comparing *average* products of specific resources on different farms or in different enterprises. However, in theory we would expect *marginal* products to be more important in determining the economic optimum level of production, combination of resources and of enterprises. If we are to use the marginal approach to decision making we must first establish the production function relating output to different levels of inputs. Hence we need a series of observations at different levels of input.

Several observations are needed even in the simplest case where there is one single variable input and one single product and the relationship is assumed to be linear (a straight line). This is illustrated graphically in Figure 13.1 showing hypothetical data relating nitrogen fertilizer input to maize yield. In Figure 13.1a we have only one single observation of input and output; there is only a single point on the graph. Obviously any number of straight lines, all with different slopes, could be made to pass through this single point. The marginal product cannot be estimated. It should be noted, however, that the average product is easily obtained by dividing output by input.

In Figure 13.1b, where there are two observations and hence two points, there is only one straight line which will pass through both. The slope and hence the marginal product per unit of nitrogen fertilizer on maize can be estimated. However, with only two points we have no way of assessing the reliability of our estimate of the slope. If there are errors of measurement of either inputs or outputs or if there are other factors influencing output which we have not taken into account, any further observations which are made may not lie close to the line at all. In order to make a more reliable estimate and to assess its reliability, many observations are needed, as in Figure 13.1c. Generally speaking, the more points that are available, the greater the reliance that can

(*a*) *Single observation*

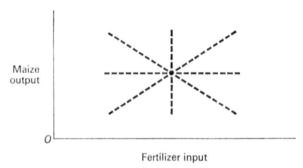

(*b*) *Two observations*

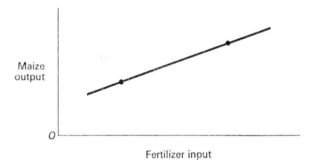

(*c*) *Many observations*

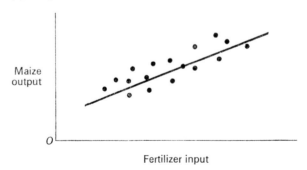

FIGURE 13.1 Fitting a straight line. Many different straight lines may pass through a single point; only one straight line can join two points, but many points are needed to provide a reliable estimate of the true relationship.

be placed on the estimated relationship. In practice the problem is further complicated where there are several variable inputs and curved relationships. More observations are needed where there are several inputs which can be varied and where various different curved relationships are considered possible.

A suitable series of observations may be obtained from controlled experiments if they were designed with this object in mind. To fit a function to experimental data, several levels of each input treatment must be included in the experiment. This is often not the case, many experiments having been designed to test whether a particular treatment, sometimes at a single level, has a 'significant effect'. However, more and more researchers in crop and livestock production are realizing the benefits of designing their experiments to measure the slope of the production response curve.

A production function may be fitted to survey data, the results for each farm representing a single observation. Various problems arise with this approach, since none of the variables are controlled as they are in an experiment. In particular, environmental conditions and managerial ability vary from one farm to another. Furthermore, since practically all inputs may vary from farm to farm some aggregation of both inputs and outputs may be needed.

A production function can only be fitted to data for a single farm, such as a unit farm, if results for several years' operation at different input levels are available. These would then represent the series of observations, in this case a time-series. Such a set of time-series data is unlikely to extend over many years so the scope for production function analysis of single farm data is limited.

METHODS OF ESTIMATING THE SLOPE

The production function may be described, as it has been in earlier chapters, by a set of tabulations of specific inputs and the related outputs. However, this is rather a clumsy method and no information is given regarding intermediate points. The economic optimum cannot be estimated precisely, but only to the nearest unit of input. It is therefore customary to attempt to relate inputs and outputs by means of a smooth curve.

Where only one input and one output is involved, the observations can be plotted on a graph as in Figure 13.1c and a curve fitted to these points by freehand drawing. Some subjective judgement is involved in drawing the shape and slope of the curve, but the accuracy must depend upon how widely the points are scat-

tered. Alternatively any one of a variety of more systematic mathematical methods can be used.

The most widely used and best known of these mathematical
techniques is the method of least squares regression. In its simplest
form, it is used to fit a straight line, such as $Y = a + bX$ in Figure
13.2, where a and b are the unknown values which are to be estimated and X is the level of input and Y the level of output. The

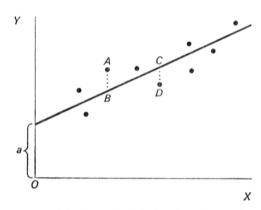

FIGURE 13.2 The straight line and deviations from it. The function $Y = a +$
bX where a is the value of Y when $X = 0$ and b is the slope of the line. The
vertical distances AB and CD are known as deviations.

object is to make the sum of the squared vertical deviations between the recorded points and the line as small as possible. The
deviations from the line are defined as the vertical distances such
as AB or CD in Figure 13.2. It will be noted that a is the value of Y
when the level of input of X is zero, and that b is the slope of the
line, and hence represents the marginal product of X.

The idea is inherently attractive. We wish to minimize deviations because this implies that the line is a good fit to our points.
If the deviations were large the line would not represent the information in a very satisfactory way. Now, if we want to make the
best possible use of our data and not waste any, every point should
be taken into account, but if all the deviations are simply added
together, large negative deviations would cancel out large positive
deviations. *Any* line which passed through the mean (i.e. the average) value of Y and the mean value of X would give a zero sum of
deviations. This is because the mean is the mid-point, so all positive deviations must exactly equal and cancel out all negative
deviations (see Figure 13.3). This problem is overcome by squaring

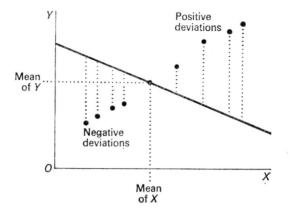

FIGURE 13.3 The sum of deviations from a line passing through the mean of X and the mean of Y is always zero. Positive deviations just balance negative deviations so the sum is always zero.

all the deviations before adding them together. Since the square of a negative real number, as well as that of a positive real number, is always positive, the sum of squares of deviations must always be positive unless the line exactly passes through all the points. Large squared negative deviations cannot offset large squared positive deviations, so the sum of squared deviations will never add up to a small number unless our line happens to fit the points closely.

It is possible to compare different lines, which have been fitted freehand, by comparing the sums of squares of deviations, *but* the mathematical technique of least squares regression enables us to estimate the specific values of a and b which minimize the sum of squares of deviations. Thus once we have decided on the general shape of the relationship between X and Y, least squares regression enables us to fit the 'best' line objectively without having to rely on our personal judgement.

Other variable inputs may be brought into the production function if we use least squares *multiple* regression. This involves successively eliminating the estimated effect of each variable and measuring the residual effect of the last one brought into the equation. Thus, if we have two variable inputs X_1 and X_2 then we can estimate the partial effect of X_2 in the following equation by estimating b_2:

$$Y = a + b_1 X_1 + b_2 X_2.$$

This is done by first eliminating the estimated effect of X_1 on both Y and X_2 by simple least squares regression.

The residual deviations in Y are then related to the residual deviations in X_2 by minimizing the resultant sum of squares of deviations. The whole process is repeated, first eliminating the effects of X_2 in order to estimate b_1, the partial effect of X_1. This procedure can be extended to cover any number of variable inputs provided that there are sufficient observations.

FORMS OF FUNCTION

I. *The linear function.* So far we have been discussing the linear function, which is based on the assumption that the inputs and outputs are all related by straight lines. This means that the slope of each relationship is constant and hence that the marginal product is constant. It makes no allowance for diminishing marginal returns so there can be no economic optimum. The total and marginal product graphs for this function are shown in Figure 13.4. It

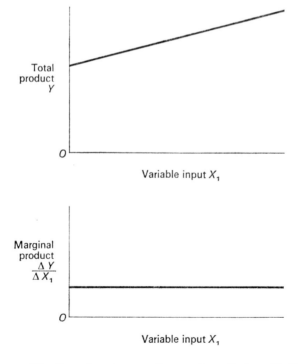

FIGURE 13.4 The linear function, e.g. $Y = 15 + 2 \cdot 5 X_1 + 3 \cdot 0 X_2$. This function represents a constant marginal product relationship so there is no economic (or technical) optimum.

is assumed that the relationships between inputs are constant (see isoquants illustrated in Figure 13.5), which means that all inputs are perfect substitutes for each other with constant rates of technical substitution. This is obviously nonsense since it would mean that the least-cost combination of resources would consist of one *single* resource input, namely the cheapest per unit of output.

Clearly the linear function is not satisfactory on theoretical grounds and its use can only be justified on the basis of the ease of fitting it by least squares regression. Fortunately, it is possible to

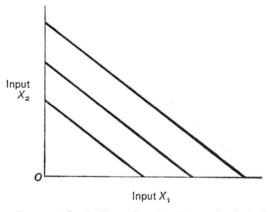

Input X_2

O

Input X_1

FIGURE 13.5 Isoquants for the linear function. Rates of technical substitution are constant so inputs must be perfect substitutes and the least-cost combination is likely to include only one input.

fit curved functions by least squares regression if the original input and output variables are transformed in some way. There are many possible transformations which may be applied but the two most common are the quadratic, in which the values of the input variables are squared, and the Cobb-Douglas in which the values of all variables are transformed into their logarithms.

II. *The quadratic function.* The equation for the quadratic function, in the case of two variable inputs, is:

$$Y = a + b_1X_1 + b_2X_2 - b_3X_1^2 - b_4X_2^2 + b_5X_1X_2. \qquad (1)$$

where Y is the level of output and X_1 and X_2 the level of each of the two variable inputs. Additional variables are formed by squaring the values of X_1 and X_2 and by forming the product of these two. Thus X_1^2 could be thought of as a new variable X_3, likewise X_2^2 could be thought of as X_4 and X_1X_2 as X_5. For instance, if for a particular observation in a fertilizer trial X_1 is 25 kg of ammonium sulphate and X_2 is 50 kg of superphosphate, then X_3 is 25^2 = 625, X_4 is 50^2 = 2500 and X_5 is 25 × 50 = 1250. The values for these new variables are calculated for other observations in the same way. The unknown values of a and all five bs can then be estimated by multiple regression of Y on the five X variables.

In the quadratic equation (1) above, b_1 and b_2 measure the direct effects of level of input on output. They are normally positive, showing a positive production response to increasing inputs of variable factor from zero upwards. On the other hand b_3 and b_4 measure the rate of change in the slope of the response curve. Thus if there are diminishing marginal returns b_3 and b_4 should have negative signs as shown in equation (1). The interaction between the two variable inputs occurs in the last term of the equation. It is usually positive, meaning that the two inputs are more productive when used in combination, but negative or zero interaction may exist where diminishing marginal returns hold true for both factors. The constant a is the output obtained when X_1 and X_2 are both zero. It therefore represents the output from the mix of fixed resources and may sometimes be zero.

The typical shapes of the total and marginal product graphs for this function are shown in Figure 13.6. This function can show diminishing marginal returns and even negative ones. There is then a technical optimum beyond which the total product falls. An economic optimum occurs where the marginal value product equals the unit factor cost. A quadratic function can show increasing marginal returns if b_4 and b_5 are positive, but it can never show both increasing marginal products at low levels of input and decreasing marginal products at higher levels of input in the same equation. Furthermore, at very high levels of input and possibly for very low ones too, this function may predict negative total products which are clearly impossible. In such a case the function ceases to be meaningful for very high or very low levels of input.

The isoquants and two possible expansion paths are shown in Figure 13.7. The isoquants are generally convex to the origin at low levels of input so they show diminishing rates of technical substitution. Thus the least-cost method of production is likely to include both variable inputs.

The isoquants may cut the axes of the graph, as is true for the lower output curve in Figure 13.7, which cuts the X_1 axis. It is implied that this level of output can be achieved by using the first variable input alone and none of the second. The expansion paths are straight lines which do not necessarily pass through zero, but converge to the point of maximum physical product. This means that the least-cost combination of resources, that is the optimum ratio of X_1 to X_2, varies according to the level of output.

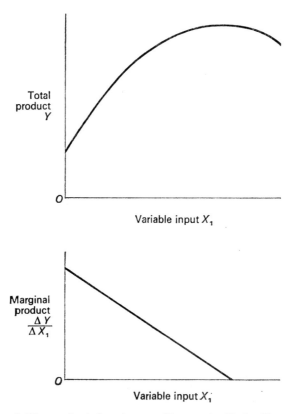

FIGURE 13.6 The quadratic function, e.g. $Y = 1 \cdot 0 + 7X_1 + 2X_2 - 0 \cdot 7X_1^2 - 0 \cdot 4X_2^2 + 0 \cdot 2X_1X_2$. This function may show diminishing marginal returns. The marginal product decreases at a constant rate down to zero at the technical optimum.

A possible disadvantage with this form of function is the large number of b values, or regression coefficients, which must be estimated for a given number of variable inputs. Thus, with a linear function the number of regression coefficients is the same as the number of variable inputs, but for a quadratic function, one variable input involves two regression coefficients, one for the X term and one for the X^2 term. Furthermore, the number of coefficients increases more rapidly than the number of variable inputs. In equation (1) with two variable inputs there are five coefficients. With three variable inputs there are ten coefficients if all possible interaction effects are estimated. Hence, if many inputs are allowed to vary in the production function the quadratic function may become very large and cumbersome and may require a very large number of observations for reliable estimation.

In summary, the advantages of this form of function are that it is relatively easy to estimate and that it may show diminishing marginal returns. The possible disadvantages are that it cannot show both increasing and diminishing marginal returns in a single

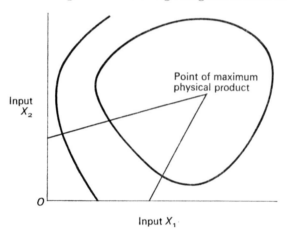

FIGURE 13.7 Isoquants for the quadratic function. Rates of technical substitution diminish, and there are many different expansion paths depending upon relative input prices, but all paths meet at the technical optimum (two are shown by thin lines).

response curve, that it may give negative total products for very high or very low levels of input and that it becomes complex if many variable inputs are included. It is most commonly used for analysing experimental results where there are few variable inputs *but* where zero variable inputs do not necessarily yield zero output.

III. *The Cobb-Douglas function.* The equation for this function, in the case of two variable inputs, is:

$$Y = A X_1^{b_1} X_2^{b_2}. \tag{2}$$

It is named after two men called Cobb and Douglas who together used it for a production function study in America in 1928.[1] The

[1] See Cobb, C. W. and Douglas, P. H. (1928) A theory of production, *American Economic Review*, 18 suppl., pp. 139–65.

a and *b* coefficients are estimated by converting all the variables measured, both inputs and outputs, into their logarithms and then using ordinary linear least squares multiple regression on these logarithms, thus:

$$\log Y = a + b_1 \log X_1 + b_2 \log X_2. \tag{3}$$

Equation (2) is simply the antilog of equation (3) so that A is a multiplicative constant and the antilog of a. The Cobb-Douglas function is sometimes known as a logarithmic function or even more precisely a double-log function, to distinguish it from other functions where only one side of the equation is transformed into logarithms. The equation is easily extended to include more variable inputs.

In this double-log equation b_1 and b_2 are direct measures of the elasticity of response for each of the input variables. Hence to arrive at the marginal product the b coefficient must be multiplied by the average product (see Appendix to Chapter 2). The average product varies however, depending on the level of input, so it is usually estimated at the average level. Where there are diminishing marginal returns, b_1 and b_2 are less than 1. A b coefficient of exactly 1 implies constant marginal returns and one greater than 1 implies increasing returns. Since the effect of scale is measured by the sum of elasticities of response for all inputs, the Cobb-Douglas function may be used to estimate returns to scale, *provided that all inputs have been included in the function*. The sum of the b coefficients then gives an estimate of returns to scale. If the sum is greater than 1 there are increasing returns and if the sum is less than 1 there are decreasing returns (see Chapter 3).

The typical shapes of the total and marginal product curves for the Cobb-Douglas function are shown in Figure 13.8. Provided that the b coefficient is less than 1, the response curve shows diminishing marginal returns. However, negative marginal returns are not possible so there is no technical optimum or maximum total product. In fact the total and marginal product curves tend to flatten out into an almost straight line at high levels of input. As a result of this, if the economic optimum occurs at a fairly high level of input, the Cobb-Douglas function may give an overestimate of the economic optimum. It will be noted that at zero level of input the output is also zero. This is invariably the case with a Cobb-Douglas function, unlike the quadratic. This is because the variable inputs in equation (2) are multiplied together, so if any one of them is zero, the product must also be zero.

The isoquants shown in Figure 13.9 are again convex to the origin, showing diminishing rates of technical substitution. They never cut the axes however, thus implying some complementarity between resources. It is impossible, according to this function, to produce any product without some of each resource; one can never

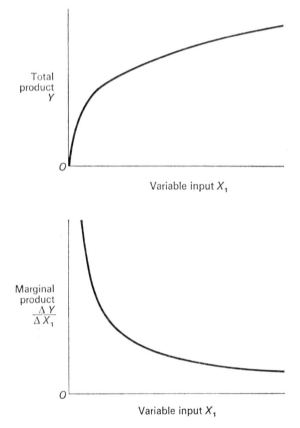

FIGURE 13.8 The Cobb-Douglas function, e.g. $Y = 2X_1^{0.3}X_2^{0.7}$. This function may show diminishing marginal returns; however, the marginal product can never fall to zero so there is no technical optimum.

substitute entirely for another. The expansion paths are straight lines passing through zero. This implies that the least-cost combination (ratio) of resources is the same at all levels of output. Once the optimum combination has been found this can be increased to scale.

The advantages of the Cobb-Douglas function are that it is easy to estimate, it may show diminishing marginal returns and can also be used to estimate returns to scale. Possible disadvantages are that it cannot show both increasing and diminishing marginal returns in a single response curve, that it does not give a technical optimum and may lead to overestimates of the economic optimum. The implication of zero output at zero input may be unacceptable in some instances. For example some crop product is usually

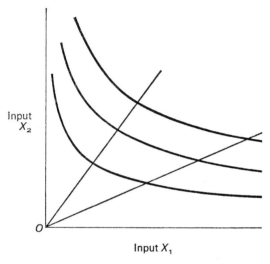

FIGURE 13.9 Isoquants for the Cobb-Douglas function. Rates of technical substitution diminish and there are many different expansion paths depending upon relative input prices but all paths are straight lines through the origin (two are shown by thin lines).

obtained even when no fertilizer is applied. The implication of a constant elasticity of response at all levels of input may also restrict the usefulness of this function. It is commonly used for analysing survey data, where many variable inputs are included and it is hoped to measure returns to scale.

PRODUCTION FUNCTIONS IN PRACTICE

The mathematical functions which have been discussed are, of course, not the only ones which could have been used to describe production relationships.[2] Nevertheless they are the ones most

[2] For other forms of function see for example Heady, E. O. and Dillon, J. L. (1961) *Agricultural production functions*, Iowa State University Press.

commonly met with in practice. The choice of function involves a certain amount of subjective judgement as does the choice of variable inputs to include in the function. It is impossible to show conclusively that one particular function is the correct one to use, since they must all be regarded as abstractions. They cannot be expected to give exact laws and there will always be some residual variation in output. It will never be possible to predict output exactly for a given level of inputs. However, as we have seen, the quadratic and the Cobb-Douglas functions have certain desirable characteristics from a theoretical point of view.

Nevertheless the frequency with which these two forms are used is probably due, in part, to the ease with which they can be estimated.

Various problems arise in applying production function analysis to any data. One general problem is that in theory the production function represents an instantaneous relationship between inputs and output. In practice there is invariably a delay between the use of the input and the output response. Thus in practice the relationship must be measured over a period of time, usually an agricultural year. However, this is not satisfactory where the delay is longer than a year, as is true for many capital investments. Furthermore, capital inputs are usually chunky or indivisible so a smooth mathematical function cannot be used to describe response. The introduction of capital inputs into a production function therefore raises considerable problems which have not been completely resolved.[3]

An associated problem is that of allowing for risk and uncertainty. The economic optimum represents the level of output which will yield the highest profit *on average*, but there must be considerable uncertainty about the outcome of any productive activity in a particular season on a particular farm. In fact, as we have seen, most farmers make some efforts to avoid risks and are willing to give up some profit for this purpose. Thus the most attractive level of output for most farmers is likely to be somewhere below the economic optimum, since they will discount the potential marginal returns for risk. Furthermore, since farmers vary in their aversion to risk, then the 'most attractive level of output' will vary between farmers. This problem also has not been completely resolved. Other problems are associated with

[3] For further discussion of this problem see Yotopoulos, P. A. (1967) From stock to flow capital inputs for agricultural production functions: a microanalytic approach, *Journal of Farm Economics*, vol. 49, no. 2, p. 476.

particular studies, whether they are 'experimental' or 'survey', so one example of each type of production function study will be discussed in more detail.

TECHNICAL EXPERIMENTATION: RESPONSE OF SORGHUM TO FERTILIZERS IN NORTHERN NIGERIA

This study was based on trials to investigate the effect of nitrogen and phosphate fertilizer on the grain yield of sorghum.[4] A total of 154 trials were carried out throughout Northern Nigeria during the five-year period 1960–4. Nine fertilizer combinations were tested in the trials, namely three levels of ammonium sulphate (nitrogen fertilizer) and three levels of superphosphate, using 0:56:112 lb. of each per acre (approximately 0:25:50 kg per hectare), in all possible combinations. At each site the nine treatments were repeated four or five times, so there were four or five observations for each fertilizer combination at every site.

It would have been possible to fit a production function to the observations for a single site but in this case the research worker made use of all 154 trials. He ended up with five response areas. Three of the five areas, namely those in the southern part of Northern Nigeria, were grouped together because the responses in each of them were too low to be of economic interest. The production functions estimated for the other two areas were:

Area I (covering most of the northern part of the country)

$$Y = 930 \cdot 58 + 77 \cdot 17 X_1 + 92 \cdot 17 X_2 + 2 \cdot 17 X_1^2 - 51 \cdot 83 X_2^2 \\ + 9 \cdot 25 X_1 X_2$$

Area II (roughly associated with the loess soils of the Zaria plain)

$$Y = 1100 \cdot 65 + 44 \cdot 67 X_1 + 223 \cdot 67 X_2 - 3 \cdot 00 X_1^2 \\ - 124 \cdot 00 X_2^2 + 45 \cdot 50 X_1 X_2$$

where Y is yield in lb. of sorghum grain per acre,
X_1 is level of nitrogen application,
X_2 is level of phosphate application.

These equations are typical quadratic functions except that for Area I, the value of X_1^2 is multiplied by a positive coefficient of $+ 2 \cdot 17$. This implies increasing marginal returns to nitrogen.

[4] See Goldsworthy, P. R. (1967) Responses of cereals to fertilizers in Northern Nigeria: 1. Sorghum, *Experimental Agriculture*, vol. 3, no. 1, p. 29; also reprinted as *Samaru Research Bulletin no. 70*, Samaru, Nigeria, Institute for Agricultural Research.

The economic optimum level of fertilizer use was not estimated since it appeared that for these two areas it would lie well above the maximum level of application used in the trials. The author rightly considered that it would be dangerous to estimate responses at levels of fertilizer use which had not been studied. However the equations were used to draw isoquants and estimate the least-cost combinations of nitrogen and phosphate fertilizers at different levels of output. Generally, the expansion path was found to lie nearly parallel to the nitrogen axis, showing that increasing pro-duct per acre is most cheaply achieved by increasing the propor-tion of nitrogen in the fertilizer combination.

This example illustrates several of the typical advantages and disadvantages of the use of experimental data for production function analysis. First there is the advantage that the inputs can be set at controlled levels so that observations may be well spaced out over the response curve and all combinations of the inputs can be investigated. This is not usually the case with survey data. Secondly, as many observations may be obtained as desired, and with a sufficiently wide range of observations it may be possible to test a number of different functions to see which of them gives the best fit. A third advantage is that most of the variables not included in the function may be held constant, for all the treatments. Thus the effects of differences in soil type can be virtually eliminated and the standard of management, the labour inputs and the timing of all operations are identical for all the different levels and combi-nations of fertilizer use.

The disadvantages of the experimental approach are associated with the fact that relatively few inputs can be varied satisfactorily in any one experiment. The majority of inputs are held constant at a specific level. This means that the interactions between the fixed inputs and the variable inputs cannot be measured, and these interactions may be very important. For instance in the work described, if the standard of management, the labour inputs or the timing of operations had been different the responses of sorghum yield to nitrogen and phosphate fertilizers might also have been different. Unfortunately the standard of management of experi-mental trials is usually much higher than that found in practice on farms.

For this reason, in general, response measured under experi-mental conditions significantly exceeds the response achieved under farm conditions. The experimental results are therefore of limited value in advising farmers or planning their farms.

An additional disadvantage of the experimental approach is that the true opportunity cost of inputs may not be known. This is because each experiment is usually concerned with only one single enterprise. Information may not be available on alternative activities. Thus in the example quoted, the fertilizers and the sorghum yield were all valued at current market prices. Now the use of fertilizer represents a short-term capital investment and it is possible that other forms of investment would yield a higher return than the market price of the fertilizer. This means that the opportunity cost of the fertilizer would be greater than its market price. Hence this type of analysis could give misleading results. In terms of market prices fertilizers might appear profitable when, in terms of opportunity cost, they might not.

FARM SURVEY ANALYSIS: SMALLHOLDER FARMING IN RHODESIA

This analysis is based on a survey of small farms on the Chiweshe Reserve in Rhodesia. Although production function analyses of farm surveys have been made in other parts of Africa[5] this one has been described and discussed in particular detail.[6]

The sample survey was originally conducted during the 1960–1 cropping season and included 118 farms in all. It is claimed that 'in terms of crop production, the 1960–1 season was approximately average for Chiweshe [and] the area sampled appears to be reasonably representative of the reserve as a whole.' A large variety of crops are grown and some livestock are kept but the analysis was restricted to the three main crops of maize (corn), millet and groundnuts (peanuts). For the production function

[5] See for instance Galletti, Baldwin and Dina, *Nigerian cocoa farmers*; Lever, B. G. (1970) *Agricultural extension in Botswana*, University of Reading, Department of Agricultural Economics, Development Study no. 7; Agabawi, K. A. and Thornton, D. S. (1967) *A study of Gendettu pump scheme, Shendi district*, University of Khartoum, Department of Rural Economy (mimeograph); Upton, *Agriculture in south-western Nigeria*, also (1968) Socio-economic survey of some farm families in Nigeria, part 2, *Bulletin of Rural Economies and Sociology*, Ibadan, vol. 2, no. 1, p. 7; and Welsch, D. E. (1965) Response to economic incentives by Abakaliki rice-farmers in Eastern Nigeria, *Journal of Farm Economics*, vol. 47, no. 4, p. 900.

[6] See Johnson, R. W. M. (1964) An economic survey of Chiweshe Reserve, *Rhodes–Livingstone Journal*, vol. 36; Massell, B. F. (1964) Farm management in peasant agriculture: an empirical study, *Food Research Institute Studies*, vol. 7, no. 2, p. 205; Massell, B. F. and Johnson, R. W. M. (1968) Economics of smallholder farming in Rhodesia, *Food Research Institute Studies*, supplement to vol. 8; and Johnson, R. W. M. (1969) The African village economy: an analytical model, *The Farm Economist*, vol. 11, no. 9, p. 359.

analysis all farms that had incomplete crop production data or that did not grow all three crops were deleted, leaving a final sample of fifty-six farms in the analysis.

A Cobb-Douglas function was used to relate the output of each crop to the set of observed inputs used in producing the crop. The function can be written, using our notation,

$$\log Y = a + b_1 \log X_1 + b_2 \log X_2 + b_3 \log X_3 + b_4 \log X_4 + b_5 \log X_5 + b_6 X_6 + b_7 X_7 + b_8 X_8$$

where Y = physical output of the particular crop (maize, millet or groundnuts),

X_1 = area of land used for that crop,

X_2 = man-hours of labour used in weeding that crop,

X_3 = weight of chemical fertilizer applied to that crop (plus a constant),

X_4 = weight of organic manure applied to that crop (plus a constant),

X_5 = value of farm implements owned, at undepreciated initial cost,

X_6 = soil type, red loam or sandy soil,

X_7 = skilled farmer, yes or no,

X_8 = semiskilled farmer, yes or no.

An extra term was added to represent the residual error but since we assume that on average the residual error is zero we can omit it from the equation.

This equation differs from the simple Cobb-Douglas function described earlier in that more variable inputs are included. Furthermore, some of the X variables do not represent quantities of a particular input, but simply take one of two values. For instance X_6 representing soil type takes the value of 1 for red loam or 0 for sandy soil. Variable X_7 takes the value of 1 for a skilled farmer or 0 for a semiskilled or unskilled farmer. Likewise variable X_8 takes the value of 1 for a semiskilled farmer or 0 for any other. Such variables are known as 'dummy variables' and are used to estimate the effects of factors which are not easily measured as physical quantities. These dummy variables are not converted into logarithms.

Before discussing the results, the variables included in the function will be considered in a little more detail.

Output is measured in physical units of weight harvested. To compare marginal value products, however, physical outputs must be multiplied by a measure of value.

Average market price paid in the area is used. For maize and groundnuts this was the official Grain Marketing Board price but for millet it was the local market price.

Land is measured in terms of the area devoted to each crop, but land is not assumed to be all of the same quality. The effect of soil type is, at least in part, allowed for by means of the dummy variable X_6.

Labour was provided by members of the farm family. For each crop, labour inputs were recorded for each of the major operations: applying manure, planting, weeding and harvesting. Because labour appeared to be a limiting factor only at weeding time, the number of weeding-hours is used as the labour variable. Hours worked by children are weighted by one-half.

Two kinds of fertilizers were used, chemical and organic, but only on the maize land. The variables X_3 and X_4 measure the quantity of fertilizer or manure input plus a constant, which in both cases is 100. A constant must be added before converting to logs since some farmers did not use any on their maize and yet still obtained some output. This implies that fertilizers are not essential inputs as there is some natural fertility in the soil. The constant may be assumed to represent the natural fertility but the choice of its value is quite arbitrary.

Fixed capital consisted of relatively simple farm implements such as an ox-drawn plough or cultivator. As an index of a farmer's fixed capital inputs, the value of farm implements at undepreciated initial cost is used. This index omits the services of draft animals and investment in the land, neither of which was recorded in the survey. Furthermore, no account is taken of the current serviceableness of the implements.

Managerial inputs are included in the function by means of the dummy variables X_7 and X_8 which are based on a rating of farmers by the government agricultural extension service. Thus farmers who receive advice from the extension service are classified into three categories: cooperators, plot holders and master farmers. A cooperator is any farmer who uses fertilizer, carries out some crop rotation and plants his crops in rows. A plot holder is a farmer who is under tuition by an extension worker to become a master farmer. A master farmer is one who has gone through the plot holder stage and has reached specified higher standards of crop and animal husbandry as laid down by the agricultural department. Of the fifty-six farms in the final sample there were three master farmers, four plot holders and fourteen cooperators. Owing to the small numbers, master farmers and plot holders are combined into a single group of 'skilled' farmers. The cooperators are referred to as 'semiskilled' and the remaining thirty-five farmers as 'unskilled'. The coefficients b_7 and b_8 are a measure of the contribution to output of 'skill' and 'semiskill' relative to lack of skill.

The results of the analysis show that there is considerable variation in output which is not explained by the production function. For groundnuts and millet less than half the total variation in output is explained by the analysis, so the authors rightly warn that the results must be interpreted with caution. The sum of the elasticities of response is nearly 1 for maize and millet, thus suggesting constant returns to scale for these two enterprises. However the sum of elasticities is only 0·753 for groundnuts, which implies decreasing returns to scale. This may be due to the omission of some important factor, such as labour quality, that should enter the groundnut production function.

The estimated marginal value products of the variable inputs from each of the three crops are given in Table 13.1, converted

TABLE 13.1 Estimated marginal value products

Input	$£$ per unit of measure		
	Maize	Groundnuts	Millet
Land per hectare	3·13	3·04	4·40
Labour per weeding hour	0·005	0·012	0·015
Chemical fertilizer per $£$ cost	1·69	—	—
Organic manure per metric ton	1·31	—	—
Fixed capital per $£$ cost	—	0·087	0·025
Soil type per hectare (advantage of red loam over sandy soil)	0·88	0·21	1·07
Skilled farmer	1·35	1·52	−1·05
Semiskilled farmer	−0·35	0·81	0·29

Note. These results are taken from Massell, *Farm management in peasant agriculture: an empirical study.* They differ slightly from those given in some of the other reports of this research.

from the original units into $£$ sterling and metric system physical measures.

These results suggest which variable inputs it is likely to be profitable to expand. There is no opportunity to bring more land under cultivation as farmers use all the arable land. Land appears to be a limiting factor. The marginal value product of labour is low in relation to the hourly wage rate in paid employment. Because of the low return to labour on the farm, many farmers spend a considerable part of the year away from the reserve working for wages. The return to chemical fertilizer does not appear adequate to justify much increase in its use. On the other hand it is suggested that the unit factor cost of organic manure is very low, virtually only the labour cost, so the marginal product is a return to labour. As an average of 16 hours was spent applying a ton of organic manure the return to this labour is about 8p per hour as against only just over 1p per hour for weeding. However, livestock numbers may be an effective constraint on the amount of manure available. The return to capital is low and the results suggest that the area is overcapitalized with respect to implements.

With regard to soil type, the benefit from farming on the red loam soils rather than sandy soils appears greatest for millet and lowest for groundnuts. The estimates for managerial skill measure technical efficiency only and do not reflect differences between farmers in allocative efficiency. For instance, the figure for skilled farmers in the maize production function is £1·35; this means that on average for a given level of all other inputs, skilled farmers obtained £1·35 more maize output than unskilled farmers. There are some unexpected results in that skilled farmers obtained *lower* returns from millet production than unskilled farmers, and the same is true for semiskilled farmers in maize production. It is suggested that this may reflect possible shortcomings in the government rating scheme which tends to focus on maize and groundnuts, or it may be the result of small sample size. The study showed that the quantities of all other resources used are related to managerial skill. Skilled farmers use more land, labour, fertilizer, manure and fixed capital than the semiskilled, who in turn use more of all these resources than the unskilled.

The analysis is also useful in suggesting how profits may be increased by reallocation of resources. The marginal value productivities of both land and labour are highest in growing millet, suggesting that profits would be raised by shifting resources from maize and groundnuts into millet production. However the resulting gain is estimated to be relatively small. In so far as the farmers may have objectives other than profit maximizing, such as self-sufficiency, the existing allocation of resources may be satisfactory. This, of course, is only true *on average*. Some individual farmers might benefit considerably from reallocation of resources.

The problems of applying production function analysis to farm survey data arise from the fact that it is impossible for the researcher to control any of the variable inputs. At least, he may restrict his study to farms of a certain size range or type, but he cannot set any inputs at selected fixed levels, or arrange that the inputs are varied independently of each other. This still need not create serious problems if it could be assumed that the quantities of inputs used on different farms varied at random. Unfortunately this is not the case, since the quantities of resources used are largely the result of conscious human decisions. For instance, if the sample of farmers all operate on the same production function, all pay the same prices for inputs and all operate at the economic optimum, then they would all use exactly the same quantity of each input and produce exactly the same output. Although there would be a large number of farms in the sample, they would all be operating at the same point on the production function. It would be impossible to draw or estimate the form of the function as in Figure 13.1a.

In fact the problem need not arise in this extreme form since some inputs are fixed at different levels on different farms, such as the supply of land, the family size, or the managerial ability of the farmer. Alternatively, there may be variation between farms in the prices paid for resources. However, it still remains true that the inputs do not vary between farms at random, but are chosen by farmers or allocated by the society according to some set of decision rules. Indeed it is likely that all inputs will vary together, as was found to be the case in the study of Chiweshe farmers just described. The skilled farmers use more land, labour, fertilizer, manure and fixed capital than the semiskilled, who in turn use more of all these resources than the unskilled. Where, as in this case, the levels of variable inputs are closely related between themselves, we speak of 'multicollinearity'. Its presence means that it is very difficult, if not impossible, to disentangle the influences of the variable inputs and obtain a reasonably precise estimate of their separate effects.

Take for instance the case of just two variable inputs, labour and land, which tend to vary together, more labour being employed on larger farms. It is then very difficult to say whether the larger output obtained on the larger farms is the result of the increased inputs of land or the increased inputs of labour, or how the extra output should be apportioned between the two inputs. We could only *safely* do this if there were some farmers who used extra land without using any more labour, that is if the inputs varied independently of each other and there was no multicollinearity present.

Where there is multicollinearity, it is particularly dangerous to omit one of the interrelated variable inputs from the function because then the marginal product of this input will be attributed to those left in the function. For example, imagine a situation where each extra hectare of land cultivated uses an extra unit of labour and yields an additional output of £26. Now this £26 represents the joint marginal product of both land and labour, but if land is left out of the production function it will appear that the £26 (or at least most of it) is the marginal product of labour alone. The omission of one of the interrelated variable inputs from the production function will therefore give overestimates of the marginal productivities of the other inputs. It will give biased results. Because of this danger it is important that *all* variable inputs should be measured and included in the production function analysis. One input which is frequently omitted, because it is difficult to measure, is the input of management, but its omission gives rise to so-called 'management bias' in the estimated marginal products of the other resources used. In the Chiweshe study, management inputs are included but only at the three levels of skilled, semiskilled and unskilled, and this ranking is based at least to some extent on the subjective judgement of the extension officers. If a more precise ranking of managerial inputs was possible, the estimates of marginal products for other inputs would probably be more accurate.

In summary it would appear that provided there is some independent variation between inputs, there is not 'exact multicollinearity' and it may be possible to estimate separate marginal products for all inputs. However, the reliability of the estimates may not be very good. Furthermore it is important that all relevant variable inputs should be included in the function to avoid bias. It may be difficult to measure some inputs such as soil quality and management and large numbers of input variables may be involved. To reduce the number of variables some attempt at grouping inputs together may be necessary. For instance, child labour is grouped together with adult labour, different kinds of fixed capital are grouped together. Such grouping must involve some arbitrary choice of which variables to group together and what relative weightings to use.

SUGGESTIONS FOR FURTHER READING

DILLON, J. L. (1968) *The analysis of response in crop and livestock production*, Oxford, Pergamon Press.

ERBYNN, K. G. (1970) *Maize response surfaces and economic optima in fertilizer use in Volta region of Ghana*, Crops Research Institute, Council for Scientific and Industrial Research, Kumasi, Ghana.

HEADY, E. O. and DILLON, J. L. (1961) *Agricultural production functions*, Iowa State University Press.

NORMAN, D. W. (1970) *An economic study of three villages in Zaria province: Part 2, Input–output relationships*, Samaru, Nigeria, Institute for Agricultural Research, Samaru Miscellaneous Paper no. 33.

OLAYIDE, S. O. and OGUNFOWORA, O. (1970) Economics of maize response to N.P.K. applications, *Bulletin of Rural Economics and Sociology*, Ibadan, vol. 5, no. 1, p. 95.

UPTON, M. (1970) Influence of management on farm production on a sample of Nigerian farms, *Farm Economist*, vol. 11, no. 12, p. 526.

Part 4 FARM PLANNING

14 Budgeting

THE NEED FOR FARM PLANNING

Agricultural development will only come about through changes in the pattern and methods of production on the individual farm units making up the whole agricultural sector. These changes may result from new market opportunities and new techniques or they may be a reallocation of the inputs already available at existing prices; these changes may occur by gradual diffusion of ideas or by a radical and rapid transformation, and they may be the result of the decisions of the many individual, indigenous producers or of direct government intervention. However, it does not follow that all changes are desirable; some will contribute to the desired development objectives, others will not. In order to choose between alternative plans, therefore, a farmer or a planner needs to estimate future outcomes. Any attempt to estimate the future outcome of a plan in quantitative terms is called a budget. Frequently the main objective is to increase the financial profit, so an ordinary budget is an estimate of the financial outcome or profitability of a particular plan or of several alternative plans. However, labour budgets are also used to estimate outcomes in terms of labour employed and capital budgets are used to predict capital requirements. Clearly, it would be possible to budget outcomes in physical units, nutritional values or any other quantitative measures provided the necessary data were available.

Planning techniques can be applied to individuals or groups of farms. The individual approach will obviously be used on a large scale unit such as a commercial plantation. Usually a highly trained manager will be employed and he may have the assistance of technical and economic advisers in drawing up plans and budgets. It is less likely that the individual approach will be used on the small farms of indigenous cultivators. As education becomes more generally available and as farmers start to keep their own records and accounts it is to be expected that they will prepare their own plans and budgets, at least to the extent of making some rough estimates on paper. However, at present there are large numbers of farmers who are unable to prepare any kind of budget. In such circumstances the cost of providing individual specialist advice and assistance would be prohibitive, so only the mass or group approach is possible.

The group approach consists of preparing plans and budgets for the typical or average farm, representative of a certain type of farm in a given area. The farm plans are then considered as the general basis for advising all farmers in the area, sometimes allowing for minor modifications to take into account individual variations in land/labour ratios, personal preferences and so on. For planning purposes the 'average farm' need not exist as a real farm. Plans can be drawn up for an abstract average or model situation without actually identifying such a farm on the ground. This can overcome the problem, already discussed, of finding a typical case study farm. If, however, the budgeted plans are to be tried out on a unit farm the problem of finding a representative farm does arise again.

The group approach may be used either on settlement schemes where government officers have a large degree of control over the patterns and methods of production, or in agricultural extension where the adviser hopes to influence the farmers by education and persuasion. Settlement schemes, both irrigated and rain-fed, are usually established on the basis of a large number of uniform sized farms, all following a standardized production plan laid down by the settlement authority. This clearly is a direct application of the group approach to farm planning and as such it should achieve rapid results, initially. However, it leaves very little scope for the development of individual initiative and planning ability among the settler farmers themselves.

Development through agricultural extension is far more dependent on the initiative and managerial ability of individual farmers. The group approach to farm planning may still prove very useful to the extension workers, providing that farms over a wide area are broadly similar in type. Although it is unlikely that every farmer would benefit from conforming to the same group plan, the planning exercise can give valuable indications as to where extension workers should concentrate their advice, which innovations are most likely to improve incomes and what are the most profitable cropping patterns to follow.

Apart from suggesting which changes and cropping patterns are likely to be most profitable to the individual farmer, farm planning and budgeting also provides the basis for evaluating their impact on the development of the total economy of the region or nation. Thus, it facilitates regional and national planning. Farm planning and budgeting is therefore an essential part of the appraisal of any proposed agricultural development programme or project.

THE BUDGETING PROCEDURE

Budgeting is really a very simple and straightforward exercise. Basically it consists of only two steps:

(1) preparing a description and specification of the proposed plan, in terms of the area of each crop and number of each class of livestock to be produced and the methods of production;

(2) estimating the expected costs and returns.

Clearly such an exercise can be carried out quickly and easily. No mathematical skill, beyond an ability to multiply, add and subtract, is needed. In practice it is a useful and flexible method of farm planning, especially suited to rapidly changing situations.

The specification of the proposed plan is based on subjective judgement or technical considerations. For instance the crop agronomist may have suggested a new crop rotation likely to improve soil fertility or permit continuous cultivation, or he may consider that a new cash crop is ecologically suited to the area; the livestock officer may consider that it would be beneficial from various technical points of view to introduce mixed farming with cattle into an area hitherto used only for crop production; the agricultural engineer may suggest that small tractors would save labour and improve the quality of cultivations. Any such proposals might form the basis of a proposed plan. In practice several alternative plans may be chosen and budgets drawn up to determine the most profitable. Some farm management economists argue that it is far safer to use budgets for comparing different plans, rather than as a guide to the profitability of a single plan.

The estimation of expected costs and returns is an exercise in forecasting the future and there is obvious scope for error. Forecasts must, of course, be based on past experience of inputs, outputs and prices. When the plan is concerned with a rearrangement of the resources and enterprises already existing on farms the best estimates of future inputs and outputs are probably the existing relationships between inputs and outputs discovered from case studies or surveys. Where a technical innovation is proposed, the input–output data must come from technical experiments. The input–output data used are in the form of averages; average resource inputs and average product outputs per hectare of crops or per head of livestock. Price expectations too are based on past prices, though this is not to say that the best estimate of next year's price is this year's price. Economists may have analysed trends in price movements which may provide a better forecast of future prices.

Ordinary financial budgets are of two basic forms, complete budgets which are concerned with the whole farm and partial budgets which show the gains or losses of a relatively minor change.

EXAMPLES OF COMPLETE FARM BUDGETS

Two examples of complete farm budgets will now be given, to illustrate the technique. Both examples were prepared for very practical use, the first as a plan for a typical farm of the Star/ Kikuyu grass ecological zone of Kenya, to be used in advising farmers, the second as a basis for one of the Nigerian farm settlements. They have been converted into metric units which has involved some slight modification of the original data but the basic methodology and assumptions are unchanged.

The example from Kenya given in Table 14.1 was drawn up by agricultural officers for a smallholding of 4·30 hectares (10·57 acres). This budget illustrates how the choice of rotation determines or at least constrains the area of each arable crop which may be grown. Thus a seven-year rotation requires seven equal sized blocks of land, each of which is laid down to a different initial course of the rotation. A course which occurs in one year out of the seven can only take up one-seventh of the total area under arable rotation. In this case the rotation also influences the area of fodder and hence the number of livestock that may be carried. This is not necessarily the best approach. Alternatively we might attempt to find an optimum combination of enterprises and then find a rotation which will fit. However, there is always a problem of organizing a cropping plan into a suitable rotation.

It is worth noting that some small scale enterprises have been omitted from the budget. There is no harm in this provided that the omission is borne in mind especially when comparing this plan

TABLE 14.1 Budget for smallholding in Kenya: Kagere Sub-location, Othaya Division, Nyeri District, Central Province

1 Specification

(a) Land use

Land use	Hectares		Hectares
Homesteads and paths	0·26	Napier grass (fodder)	0·36
Arable rotation	1·54	Vegetables	
Coffee	0·40	(home consumption)	0·23
Napier grass (mulch)	0·42	Bananas (home consumption)	0·23
Cassava (home consumption)	0·26		
Permanent grass and trees	0·60	Total	4·30

(b) Arable land

Seven course rotation on 1·54 hectares as follows:

	Long rains	Short rains
First year	Grass	English potatoes
Second year	Beans	Late maize
Third year	English potatoes	Beans
Fourth year	Early maize	English potatoes
Fifth year	Early maize	Grass
Sixth year	Grass	Grass
Seventh year	Grass	Grass

The land is thus cropped for 4 years and rested under pasture for 3 years. The area under each course must be 1·54/7 = 0·22 hectares. Hence the area of each of these crops is:

	Hectares		Hectares
Grass (3 courses = 3 × 0·22)	0·66	Early maize (2 courses)	0·44
English potatoes (3 courses)	0·66	Late maize (1 course)	0·22
Beans (2 courses = 2 × 0·22)	0·44		

(c) Livestock

Dairy cows to be kept on available fodder area at a stocking rate of 1 adult beast per 0·4 hectares.

Types of fodder	Hectares
Rotational grass	0·66
Napier grass (fodder)	0·36
Permanent grass and trees	0·60
Total area of fodder	1·62

Hence: can carry 1·62/0·4 = 4 dairy cows.

TABLE 14.1—*continued*

(*d*) Cassava, vegetables and bananas used for home consumption—looked upon as private garden, sparetime activities, therefore excluded from the budget. Hence commercial products limited to english potatoes, beans, maize (early and late), coffee and milk.

(*e*) Family labour force plus hired oxen.

2 Estimated returns and costs (East African shillings)

Enterprise	Number of units	Price of product	Yield	Gross output	Costs	Net return	Total net return
			per unit				
	hectares	*s.* per bag	bags	*s.*	*s.*	*s.*	*s.*
English potatoes	0·66	13	98	1274	189	1085	716
Beans	0·44	43·25	15	648·75	32·75	616	271
Early maize	0·44	25	25	625	15	610	268
Late maize	0·22	25	20	500	14	486	107
		s. per kg cherry	kg parchment				
Coffee*	0·40	1·1	1253	9511	11	9500	3800
		per cow					
	cows	*s.* per litre	litres	*s.*	*s.*	*s.*	
Milk	4	0·33	910	300	50	250	1000

Total		6162
Less cost of oxen hire		140
Net income (return on capital, management and labour)		6022

*Parchment: cherry ratio 1:6·9.

After Clayton, E. S. (1961) Economic and technical optima in peasant agriculture, *Journal of Agricultural Economics*, vol. 14, no. 3, p. 337 (reprinted in Whetham and Currie, *Readings in the Applied Economics of Africa*).

with others. In estimating the costs of the plan, land has been omitted presumably because it is costless. Labour has also been omitted from costs so that the final net income represents the return to labour as well as profit. No charge has been made for return on capital invested in coffee trees and livestock, land improvements and so on, so this return must also be included in the net income. All costs other than oxen hire charges have been treated as direct costs and allocated to individual enterprises. In fact these direct costs consist only of seed in the case of arable crops, depreciation of trees and sprays for tree crops, and depreciation and spraying costs for livestock, so the net returns per unit of each enterprise are really gross margins.

The second example (from Nigeria) given in Table 14.2 was drawn up by the F.A.O. advisory team on farm settlements in Nigeria for a much larger farm of 28 hectares. Although prepared for different conditions in a different area, the budgeting procedure is quite similar to that used in the Kenya budget. There was possibly less *local* information or experience available in this case as some of the estimates are admitted to be unreliable. In particular no details are available as to how the quite large return per hectare of miscellaneous crops is made up. Also, the assumption that grazing cattle will exactly break even making neither profit or loss is a reflection of the uncertainty regarding this enterprise.

The dangers of double counting or of omission are illustrated by this budget. Although the maize and sorghum are to be fed to the poultry, thus saving on poultry feed costs, for the purposes of this budget the value of the maize and sorghum has been charged against the poultry. If this had not been done the returns from maize and sorghum would have been counted twice, first as the output of these crops and second as the reduction in poultry food cost. Alternatively double counting might have been avoided by reducing the poultry feed cost but not recording the output of maize and sorghum. There would then have been a danger of forgetting the costs of producing the maize and sorghum.

However, neither of these budgets is presented as a particularly reliable estimate of future performance. They are simply given as typical examples of the technique. Clearly the results are unlikely to be applicable in other areas, or in other time periods than where and when they were prepared.

ADVANTAGES AND LIMITATIONS OF THE BUDGETING TECHNIQUE

Budgeting in the form just illustrated is not the only technique of planning for the future. Production function analysis is a planning technique, in that it may be used to choose between alternative enterprises and methods of production. Linear programming is another method of drawing up future plans. Furthermore these

TABLE 14.2 Budget for settlement farm in Nigeria: arable land settlement, Western Region

1 Specification

(a) Land use	Hectares
Arable rotation	20
Cashew plantation	4
Teak plantation for poles and timber	4
Total area per settler	28

(b) Arable rotation

Eight course rotation on 20 hectares as follows:

	Long rains	Short rains
First year	Early maize	Dwarf sorghum
Second year	Early maize	Cowpeas
Third year	Early maize	Dwarf sorghum
Fourth year	Early maize	Cowpeas
Fifth to eighth year	Grass legume pasture	

The land is thus cropped for 4 years and rested under pasture for 4 years. The area under each course must be $20/8 = 2 \cdot 5$ hectares.

However, in addition, 1 hectare of rotational land is to be used for miscellaneous cropping (intensive vegetable production) each year. For this purpose 1 hectare is taken out of early maize production and 0·5 hectare out of both dwarf sorghum and cowpea production. Hence the area of each arable crop is:

	Hectares
Pasture (4 courses $= 4 \times 2 \cdot 5$ hectares)	10
Early maize (4 courses $=$ 10 hectares less 1 for miscellaneous crops)	9
Dwarf sorghum (2 courses $= 2 \times 2 \cdot 5 = 5$ hectares less 0·5 for miscellaneous crops)	4·5
Cowpeas (2 courses $= 5$ hectares less 0·5)	4·5
Miscellaneous crops	1

(c) Livestock

Poultry unit of 192 laying hens. Cattle to be grazed on the pasture—but the net return from cattle is taken to be nil.

(d) Family labour force plus tractor hire. Tractor hire charges included in the costs of each enterprise.

(e) Land to be held under perpetual lease with fixed nominal rent per hectare. Also included in the estimated costs.

TABLE 14.2—*continued*

2 Estimated returns and costs (Nigerian pounds)

Enterprise	Number of units	Price of product	Yield	Gross output	Costs	Net return	Total net return
			per unit				
	hectares	£ per quintal	quintals	£	£	£	£
Maize	9	2	22	44	31	13	117
Sorghum	4·5	2	18	36	29·20	6·80	30·60
Cowpeas	4·5	3	5	15	6·50	8·50	38·25
Miscellaneous crops	1	—	—	—	—	60	60
Cashew nuts	4	5·50	8	44	11·50	32·50	130
Teak	4	—	—	—	—	12·50	50
			per hen				
	hens	£ per dozen	eggs	£	£	£	
Poultry	192	0·20	180	3	2·35	0·65	125

Estimated annual return to labour and management per settler	550·85

Source: Agrawal, G. D. (1964) *Farm planning and management manual,* Ministry of Agriculture and Natural Resources, Western Nigeria.

other approaches have the advantage over budgeting that they lead in a systematic way to an optimal, or most profitable, solution. Thus a production function analysis or a linear programme gives results in terms of the specific combinations of enterprises and levels of resource use which will give a maximum profit. In comparison, budgeting appears to be a technique of trial and error. It can be used to find the most profitable of, say, three alternative plans but the budgeting process does not guide the original choice of plans. Whether the chosen plan is near the best possible plan depends entirely on whether one of the original three alternatives is near the best possible solution. Of course some subjective judgement is involved in choosing the enterprises and resources to be considered for other methods of planning, but budgeting involves rather *more* subjective judgement.

It is usually recommended that economic principles should be borne in mind in drawing up budgets. The possibilities of diminishing marginal returns or increasing marginal costs, of supplementary or complementary relationships between enterprises and resources and discontinuous or lumpy inputs should all be considered. However, the technique of budgeting does not ensure that they are considered and, in practice, diminishing marginal returns are often ignored, constant average costs and returns being used.

However, the major advantage of budgeting, it is worth repeating, is its simplicity and flexibility. Practically any data can be used, ranging from near guesses to information which has been very accurately collected and analysed. Furthermore the necessary data are limited to standard average inputs and outputs, unlike production function analysis which requires many input–output combinations to predict marginal returns. In fact the necessary data for production function estimation are rarely available. For this reason the production function approach can only be used in a limited number of cases where the necessary information has been collected. Since budgeting requires far less data of a less detailed nature, and because of its simplicity, it will always have a place in practical farm planning.

LABOUR BUDGETS

Most authorities refer to the need for listing the available resources on each farm, including areas of land of different classes, available labour force, capital resources and the attitudes and managerial ability of the farmer. The purpose of this listing is to ensure that the budgeted plans are feasible, that is, that they do not require more resources than are likely to be available. To make the feasibility check, then, it is necessary to estimate the resource requirement to compare with the quantities available.

An estimate of the labour requirement of a farm plan is known as a labour budget. In its simplest form it involves the calculation of a total standard man-day requirement for the plan. This total requirement is calculated by multiplying the number of hectares of each crop and the number in each class of livestock by average *annual* requirements per hectare and per head of stock. An obvious difficulty arises due to the different seasonality of various types of work. The total seasonal labour requirement at any time depends upon the way in which the seasonal labour requirements for individual enterprises come together. As noted in Chapter 7 on labour, it is the seasonal work peaks which determine the regular labour needs.

Thus we really need to estimate the labour requirement month by month (or week by week or at some other suitable interval). For this purpose standard monthly requirements per hectare of crops and per head of livestock are needed. The monthly total requirements can then be plotted on a graph as in Figure 7.5 (page 141) to show the labour profile. Incidentally from the labour profile Clayton was able to show that the original budget prepared for the farm by the agricultural advisers was not feasible with the available labour force. Even where labour can be hired on a temporary basis it is useful to estimate when and how much of it will be needed. Hence it is always advisable to prepare a labour budget for any farm plan.

CAPITAL BUDGETS AND PHASING

The estimation of capital requirements by means of a capital budget raises various problems. A simple version is set out in Table 14.3 showing the estimated capital requirement for the Nigerian settlement farm described in Table 14.2. Costs of roads, housing and social facilities are not included.

TABLE 14.3 Capital budget for settlement farm in Nigeria

	£
Establishment cost of cashew plantation	133
Establishment cost of teak plantation	122
Poultry housing	320
Stock: 240 day-old hen chicks (allowing for losses)	36
Share of tractors and machinery	800
Share of grain dryer/store	377
Total productive capital per settler	1788

This budget is oversimplified since it makes no allowance for working capital. Thus additional capital will be required to feed the chicks up to the point of lay, to keep the farmer and his family and to pay labour and other expenses which arise before the first harvest. Furthermore the costs of establishing the cashew and teak plantations will not be recovered for several years. The total requirement, of both fixed and working capital, must depend upon the phasing or timing of the steps in the development of the plan. Where costs are incurred before there is any revenue to meet these costs, capital is needed. However, where sufficient revenue is obtained before costs are incurred then extra capital will not be needed.

This may be illustrated by assuming a particular phasing for the plan budgeted in Tables 14.2 and 14.3. The expected monthly cash flows are given in Table 14.4. The cumulative deficit gives a

TABLE 14.4 Possible capital profile for arable settlement farm in Nigeria

Month	Notes	£		Net cash flow	Cumulative deficit
		Expenses	Revenue		
January	Settlement started: cost of poultry housing, chicks, feed, tractors and machinery, family subsistence	1184	nil	−1184	−1184
February	Clearing for trees, poultry feed, subsistence	124	nil	−124	−1308
March	Grain dryer, maize established, poultry feed, subsistence	680	nil	−680	−1988
April	Trees, poultry feed, subsistence	124	nil	−124	−2112
May	Poultry feed, subsistence	24	nil	−24	−2136
June	Poultry feed, subsistence	24	nil	−24	−2160
July	Poultry feed, subsistence	24	nil	−24	−2184
August	Poultry feed, subsistence, sorghum and cowpea costs, sales of maize	184	396	212	−1972
September	Poultry feed, subsistence	24	nil	−24	−1996

profile of the current capital requirement. The maximum value or peak determines the overall capital requirement (see Chapter 8).

According to this phasing of the plan, the peak capital requirement will occur in July and is considerably more than the original estimate in Table 14.3. It might be necessary to carry the estimates on even into the second or later years of the programme if there is any likelihood of a peak higher than £2184 occurring. In fact, the high initial costs of machinery, equipment and poultry housing almost certainly will cause the peak to occur in the first year in this example. However, if the machinery and poultry are gradually phased into the plan over a period of years the peak capital requirement would probably be lower because earnings from crop sales could contribute towards the investment. The peak capital requirement might also be delayed for a few years after the start of the settlement.[1]

With tree crops there is inevitably a delay of some years before the peak capital requirement occurs. Thus Table 8.1 (page 155) is a capital profile for a hectare of oil palms. Where such a long delay occurs before the peak capital need arises there is the additional cost of waiting which may take the form of actual interest charges. Then, the techniques of compounding or discounting described in the appendix to Chapter 8 should be used. The correct capital profile for the oil palms is really shown in Figure 8.4 rather than Table 8.1.

In fact the preparation of a phased plan and a profile of cash flows is necessary if discounting is to be used for appraisal of any proposed investment plan. Thus not only does the capital budget show when the peak requirement occurs and how big it is but it also permits the calculation of the internal rate of return or the net present value of the proposed plan. Clearly capital budgeting is essential in planning development projects for which the government or any other outside body is providing capital. This outside body needs to have a precise estimate of the total capital requirement and the returns which may be expected in order to decide whether the project is worth financing. Usually there are many alternative projects competing for the limited development funds available and a choice must be made between them. Capital budgets and estimates of the likely returns are then needed to rank them in order of priority.

Many agricultural credit schemes have been launched with very little planning. As a result loans are often made to farmers according to their credit standing or their possession of wealth, and not

---

[1] See Ogunfowora, O. and Heady, E. O. (1972) An integration of short-term farm enterprises with perennial tree crops: an application of recursive programming to a tree crop farm settlement in Western Nigeria, Journal Paper no. J7069 of the Iowa Agricultural and Home Economics Experiment Station, Project 1558, submitted to *Journal of Developing Areas*, Illinois.

according to the possible contribution of the loan to the farmers' earning power and welfare. Using the group approach, capital budgets for a typical farm will suggest ways in which credit can be used most profitably, how much is needed and what returns are likely to be obtained. A rational and effective credit system, therefore, should be arranged to meet the farmer's conditions and needs as they are determined through farm planning and budgeting.

BUDGETARY CONTROL

Budgets may be used not only for farm planning but in some circumstances for control of the business too. Where records are kept they may be checked regularly against the budgeted plan. This is particularly likely to be feasible on a government-run plantation or a settlement scheme. However, it does require that budgets are prepared and records are kept in some detail. The budgets must specify physical quantities of inputs and outputs as well as expected costs and returns.

In order to keep tight control, timing is very important. Thus it is not very helpful to discover at the end of the year that poultry feed consumption was twice the budgeted figure. In order to correct the fault, the farmer needs to know as soon as possible after it occurs. He needs to know within a week or less if poultry are being fed too much. Hence for efficient budgetary control the timing and phasing of inputs and outputs must be set out as in a capital profile estimation.

When actual results differ from those planned it is sometimes difficult to decide whether it is the original budget or the management which is at fault and in need of correction. Nevertheless it is valuable to pinpoint where differences occur. Further checks can then be carried out to decide where the fault lies.

SUGGESTIONS FOR FURTHER READING

COLE, J. C. (1968) *Budgeting and budgetary control*, University of Newcastle on Tyne, Agricultural Adjustment Unit, Technical Paper no. 6.

GILES, A. K. (1964) *Budgetary control as an aid to farm management*, University of Reading, Department of Agricultural Economics, Miscellaneous Studies no. 33.

STURROCK, F. G. (1971) *Farm accounting and management*, 6th ed., London, Pitman.

15 Partial budgets and programme planning

Where only a relatively minor change in the pattern of farming is proposed it is not necessary to prepare a complete farm budget to estimate the result. Instead a partial budget can be used to arrive at the expected *change* in profits. This takes into account only those changes in costs and returns that result directly from the proposed modification. Farm costs or returns which are unaffected by the proposal are excluded from the calculation.

The simplest form of partial budget, applicable where a new enterprise or a new process such as crop spraying is introduced, involves the following questions:

(*a*) what extra returns (gains) can be expected?
(*b*) what extra costs will be incurred?

Where the proposed new activity substitutes for something already existing, as when one crop substitutes for another or a machine substitutes for labour, we must also ask:

(*c*) what present costs will no longer be incurred?
(*d*) what present income will be sacrificed?

Hence the total gain will be (*a*) + (*c*), the extra returns plus the saved costs, and the total cost will be (*b*) + (*d*), the extra costs plus the present income forgone. The total gain minus the total cost then represents the net gain or expected increase in profit.

As with complete budgets, the first step in partial budgeting should be a description and specification of the proposed change stating clearly what is involved and when it occurs. Information should be included on the stock numbers, the areas of crops and the methods of production to be used. As a second step in partial budgeting it is useful to list those items in the existing system likely to be changed when the new policy is introduced. This reduces the likelihood of omitting possible indirect effects of the change. Appropriate values can then be used to predict extra returns, savings in present costs, extra costs and income sacrificed to arrive at the expected change in profit.

A partial budget can be compiled more quickly and easily than a complete budget, since it is only concerned with the costs and returns that are to be changed. The many items of cost and income unaffected by the change need not be estimated. However, when a major change is made, such as the introduction of a tractor in place of hand labour, this will influence most of the existing inputs and outputs; the whole pattern of farming will be modified. In such circumstances a partial budget is unsuitable and a complete budget is needed.

All that was said regarding feasibility testing and the need for labour and capital budgets in connection with complete budgeting, applies equally to partial budgets. However, with a partial budget it is only necessary to estimate the change in labour or capital requirements. Strictly speaking Table 8.1 is a partial capital budget showing the *extra* capital requirement of growing a hectare of oil palms.

EXAMPLES OF PARTIAL BUDGETS

For the first example we will consider the effect of introducing grain storage, where hitherto farmers had been forced to sell their guinea corn (sorghum) at harvest-time. The price is expected to increase from £20 per ton at harvest-time to £24·25 per ton after 9 months' storage. This may not be a very realistic situation and the figures used are hypothetical but it serves to illustrate the technique. The results are given in Table 15.1.

This proposal represents a capital investment, so return on capital should be the main criterion for deciding whether to adopt the plan. Ideally a capital profile should be prepared by estimating cash flows for successive three-monthly or six-monthly periods over the whole life of the storage bin. Depreciation charges would be omitted. These cash flows could then be used to estimate the internal rate of return on capital. However, this would be a complex task, especially as there might be interactions with other enterprises. Thus sales of groundnuts might provide the working capital necessary to maintain the family and farm until the guinea corn was sold. A complete budget and capital profile would therefore be needed. Hence the less accurate approach shown in Table 15.1 is used to give an estimate of the net gain. This may be compared

TABLE 15.1 Partial budget to estimate extra profit from storage

---

1 Specification

Proposal to store 10 tons (10,000 kg) sorghum per year. Cost of small concrete bin of 10 tons capacity £100. Expected life of bin 10 years. No maintenance or repair cost. Grain to be fumigated, so losses due to insect damage and drying in the bin negligible.

---

2 Items in present system likely to be changed

Sorghum no longer available for sale at harvest-time. No reduction in existing costs.

---

3 Estimated gains and costs (£)

Gains		Costs	
(a) Extra returns		(b) Extra costs	
Sales of stored grain		Cleaning bin	0·25
10 tons at £24·25	242·50	Fumigation of bin	
		15p per ton	1·50
		Depreciation of bin	
		£100/10	10
(c) Saved costs	nil	(d) Present income sacrificed	
		Sales of grain at harvest-time (10 tons at £20)	200
Total	242·50		211·75

Net gain = 242·50 − 211·75 = 30·75

TABLE 15.2 Capital budget for proposal in Table 15.1

	£
Fixed capital	
Initial cost of bin	100
Working capital	
Income forgone by not selling sorghum at harvest-time	200
Cost of cleaning bin	0·25
Cost of fumigation	1·50
Total initial capital	301·75

Return on initial capital = 30·75/301·75 × 100 = 10%

with the initial investment as in Table 15.2 to give a rough idea of the return on this initial investment.

The second example concerns a change in cropping pattern, namely the substitution of a hectare of cotton for a hectare of groundnuts. Again hypothetical figures are used. In this case total gains will consist of new income from the sale of cotton, and the saved costs of groundnut production. Total costs will include both the new costs of producing cotton, and groundnut income sacrificed. The partial budget is set out in Table 15.3.

TABLE 15.3 Partial budget to estimate net gain from substituting cotton for groundnuts

---

1 Specification

One hectare of cotton to be substituted for 1 hectare of groundnuts.

---

2 Items in present system likely to be changed

    (*a*) Loss of output of 1 hectare of groundnuts
    (*b*) Saving in seed cost for 1 hectare of groundnuts
    (*c*) No change in total land use
    (*d*) No change in total labour force
    (*e*) No change in capital requirements

---

3 Estimated gains and costs ($\pounds$)

Gains		Costs	
(*a*) Extra returns		(*b*) Extra costs	
Cotton production		Cotton seed	
1200 kg at 5p per kg	60	12 kg at 8p per kg	1
		Fertilizer	
		200 kg superphosphate	
		at $\pounds$1·50 per 100 kg	3
		200 kg sulphate of	
		ammonia at $\pounds$2 per 100 kg	4
		Spray	
		20 litres at 10p per litre	2
(*c*) Saved costs		(*d*) Present income sacrificed	
Groundnut seed		Groundnut production	
133 kg at 6p per kg	8	800 kg at 6p per kg	48
Total	68		58

Net gain = 68 − 58 = 10

USE OF GROSS MARGINS

The budget set out in Table 15.3 is typical of all budgets concerned with relatively small changes in the pattern of enterprises making up the farm system. The common characteristic of such changes is that they do not alter the fixed costs of land, labour and fixed capital. Only variable costs are affected by the change. This implies that the effects of the change could be estimated by using gross margins, and this is generally the case. Thus Table 15.3 could be rearranged by subtracting the extra (variable) costs of cotton from the extra returns to give the gross margin for cotton, and subtracting the saved (variable) costs of groundnuts from the present groundnut income forgone to give the gross margin for groundnuts. The net gain from substituting a hectare of cotton for a hectare of groundnuts is now the extra gross margin. This is set out in Table 15.4.

TABLE 15.4 Estimated gains and costs of substituting cotton for groundnut production in terms of gross margins

Gains (£)		Costs (£)	
Extra production from introducing cotton		Groundnut production forgone	
1200 kg at 5p per kg	60	800 kg at 6p per kg	48
Less extra costs		*Less* costs saved	
Seed	1	Seed	8
Superphosphate	3		
Sulphate of ammonia	4		
Sprays	2		
Equals gross margin	50	Equals gross margin	40

This means that if we know the gross margins of the two enterprises it is a very quick and easy exercise to estimate the result of substituting a hectare of cotton for a hectare of groundnuts or even of substituting 0·6 hectares of cotton for 1·1 hectares of groundnuts or any other relative change that might be feasible.

Consideration of the feasibility of the change now suggests a way in which the plan can be selected on a more rational basis. That is, if we can discover which resource is the most limiting constraint on production, we can select the enterprise which yields the highest gross margin per unit of that limiting constraint. Furthermore, we can decide objectively how far to expand that enterprise. In fact we should expand it as far as possible until all the limiting resource is used up.

To illustrate this we will assume that labour is the resource which limits production. Hypothetical labour requirements per hectare of the two crops budgeted in Tables 15.3 and 15.4 are set out in Table 15.5.

TABLE 15.5 Monthly labour requirements per hectare of cotton and groundnuts

	Man-days	
	Cotton	Groundnuts
January	10	14
February	8	—
March	—	—
April	12	—
May	22	—
June	24	16
July	17	12
August	17	8
September	—	7
October	—	8
November	15	15
December	25	20
Total	150	100

Now if labour is only available on a regular basis throughout the year, as is probably true of family labour, then the peak require-ment must limit the area of crop that can be grown. In this case the peak requirement for both crops occurs in December, when cotton requires 25 man-days per hectare and groundnuts 20 man-days. Hence we should choose the crop which yields the highest gross margin per man-day of labour in December. For cotton the gross margin per hectare is £50 so the gross margin per man-day of December labour is £50/25 = £2. For groundnuts the compar-able figure is £40/20 = £2. Hence there is nothing to choose between the two crops. However, if we assume that some of the December work might be carried out in November, to spread the peak a little, then we should compare the gross margin per man-day required in November and December together. For cotton the total labour requirement for the two months is 40 man-days and for groundnuts it is 35. Now the gross margin per man-day over the two months is for cotton £50/40 = £1·25 and for groundnuts £40/35 = £1·14. Hence cotton is the more attractive alternative. The same is true if we take the December and January labour requirements together.

Now if we know just how much labour will be available in each month we can calculate the amount of cotton that may be grown. Let us say that a surplus of 20 man-days is available in each month for growing either cotton or groundnuts. If the December work load could not be spread into November then the maximum possible area of cotton would be 20/25 = 0·8 hectares. This would yield a gross margin of £50 × 0·8 = £40. However if we assume that it can be spread equally between the two months the labour available would be 40 man-days and the labour requirement per hectare 15 + 25 = 40. Hence a whole hectare could be grown, yielding a gross margin of £50.

Using the same hypothetical labour requirements and gross margins we may consider an alternative situation where casual labour can be hired as and when needed. The main constraint on the total amount of labour available might then be the supply of capital. For example let us suppose that the farmer can afford to hire just 300 man-days of labour but at any time of the year, the daily wage-rate being constant. Now he should choose the enterprise which yields the highest gross margin per man-day of total labour requirement. Cotton needs 150 man-days per year so the gross margin per man-day is £50/150= £0·33. Groundnuts, however, require only 100 man-days so the gross margin per man-day is £40/100 = £0·40. Hence in this case, provided that the cost of hiring labour is less than £0·40 per man-day, groundnuts should be produced in order to maximize profits.

The total area of groundnuts that can be grown with this additional labour force is estimated by dividing the total number of man-days available, namely 300, by the labour requirement per hectare of 100 man-days. Hence the maximum feasible area of groundnuts in 300/100 = 3 hectares, which will yield a total gross margin of £40 × 3 = £120.

This approach of selecting enterprises to introduce or expand according to the gross margin per unit of limiting resource is systematic and logical. So too is the policy of expanding that enterprise until the limiting resource is all used up. The major problem which may arise in practice is that of discovering which resource is likely to be the most limiting. In this respect the above examples were highly simplified. Thus it was particularly convenient that the peak labour requirement for both crops occurred in the same month, namely December. Had the peak requirement for cotton been in June for example, it would be difficult to decide which month was to be treated as the most limiting. The problem would be much more complicated if the man-days of labour available varied from month to month. If, in December, there was more labour available than in other months, then December labour might no longer be the limiting constraint.

In these more complex situations the most limiting resource may be found by estimating the maximum gross margin which can be earned by using each resource up to its limit. The lowest value occurs where the resource is most limiting. This may be illustrated still using the comparative labour requirements and gross margins for cotton and groundnuts. Ignoring other inputs of land or capital which might act as constraints, and ignoring possible social and cultural constraints, we consider only labour availability. It is assumed to vary from month to month as set out in Table 15.6.

TABLE 15.6 Monthly labour available (man-days)

January	28	May	20	September	25
February	16	June	24	October	22
March	16	July	24	November	30
April	24	August	24	December	47

Total for year 300

Now we must consider the gross margin per man-day which may be earned by each enterprise in each month. This is obtained by dividing the labour requirement in each month from Table 15.5 into the gross margin per hectare. Thus the gross margin per man-day of January labour in cotton production is £50/10 = £5. The results for all months are set out in Table 15.7.

The gross margin per man-day for each month is now multiplied by the number of man-days available to arrive at the gross margin which would be obtained if all the labour in that month could be allocated to that particular crop. Thus since the gross margin per man-day of January labour is £5, and since there are 28 man-days available in January, if this could all be used in cotton production the total gross margin would be £5 × 28 = £140. The results for all months for both crops are set out in Table 15.8.

TABLE 15.7 Gross margin per man-day of monthly labour

| | £ | |
	Cotton	Groundnuts
January	5	2·86
February	6·25	∞
March	∞	∞
April	4·17	∞
May	2·27	∞
June	2·08	2·50
July	2·94	3·33
August	2·94	5
September	∞	5·71
October	∞	5
November	3·33	2·67
December	2	2
For whole year	0·33	0·40

Note. The infinity sign ∞ implies that since no labour is required in that month the gross margin per man-day is infinitely large.

The smallest value in each row now represents the maximum gross margin that can be earned from that crop because it is limited by the labour supply in that month. Thus the smallest value for cotton occurs in May when the total gross margin is £45.

TABLE 15.8 Total gross margins possible with specialization

| | £ | |
	Cotton	Groundnuts
January	140	80
February	100	∞
March	∞	∞
April	100	∞
May	45	∞
June	50	60
July	71	80
August	71	120
September	∞	143
October	∞	110
November	100	80
December	94	94
For whole year	100	120

Although there is sufficient labour in every other month to produce a higher gross margin from cotton, the May labour supply limits the amount that can be produced and hence the gross margin that can be obtained. May labour is the most limiting constraint for cotton production. The smallest value for groundnuts however occurs in the June column, so June labour is the limiting constraint for groundnuts.

Now to choose between the two enterprises we compare the maximum gross margin that can be earned from cotton production with the maximum gross margin that can be earned from groundnuts. Hence we compare the gross margin obtainable when all May labour is used in cotton production, namely £45, with that obtainable when all June labour is used in producing groundnuts, namely £60. Thus for the pattern of labour supply set out in Table 15.6 and the gross margins and labour requirements for the two crops as given it would appear that the most profitable choice is groundnuts. The area which may be grown is determined by dividing the man-days available in June by the labour requirement in June, which gives $24/16 = 1.5$ hectares.

It may be noted that the critical or most limiting constraint could have been identified by calculating the area of each crop which might be grown with the given labour supply in each month. The smallest value for each crop would then represent the maximum possible area and the month in which it occurred would represent the critical labour constraint. To compare the two crops the maximum possible area of each would be multiplied by the relevant gross margin per hectare. Naturally, this should give the same result as the first approach.

This method of planning changes is systematic and logical, but clearly it is laborious and time consuming. In this connection it should be noted that most real life problems involve more alternative enterprises and more constraints.

PROGRAMME PLANNING: AN EXAMPLE

Partial budgeting, as we have already noted, is used only for estimating the outcome of relatively small changes in an existing farm system. We can assume that many items of cost are unlikely to be affected by such changes, so these items can be ignored in preparing partial budgets. When budgeting changes in the combination of enterprises produced we need only compare gross margins as shown above.

However, when it comes to planning the whole farm we are often faced with the fact that the farmer has a given fixed supply of certain resources which he wishes to allocate in the most profitable manner. Thus his labour supply may be restricted to family members and his capital resources limited to what he can save. How then should he allocate these limited resources? A solution may be found by extending the gross margin approach, just described, to planning the whole farm. Any method of whole farm planning which relies on gross margins other than the completely systematized method of linear programming is usually called 'programme planning'.

It will be noticed that the basic assumption made here differs from the assumptions used in production function theory and analysis. The assumption here is that diminishing returns do not occur as more and more variable resource is used in a particular enterprise, that is to say that the gross margin per hectare remains constant no matter how many hectares of a particular crop are produced. The ultimate size of each enterprise is then limited by the fixed supply of certain resources. In production function analysis it is generally assumed that most resources are variable and the size of each enterprise is established at the economic optimum.

Programme planning was used to plan the unit farm at Ukiriguru, Tanzania, at yearly intervals.[1] For the first season there was no disposable income generated from the farm itself. Available capital was limited to the credit facilities available in the area. This restricted purchased inputs to fertilizer for the cotton crop, except that seeds were also provided on credit where the farmer could not provide his own. One important implication of this limitation placed on credit was that the farm had to be worked by family labour only for this first season. There could be no casual hired labour.

The first constraint on profit maximization was assumed to be the need to provide for the family's subsistence needs from the farm. This is not a resource constraint but is a real constraint imposed by local custom and a desire for security. The farmer himself decided which food crops he wanted for his family. The only condition was that at least one each of grain, legume and root crops should be included as all three are grown on most farms in the area. For the first season he selected maize, rice, sweet potatoes,

---

[1] Collinson, *Experience with a trial management farm in Tanzania.*

cassava and groundnuts. Estimates were then made of the quantities of each crop needed to provide a balanced diet for the family, standard nutritional requirements being used as a guide. Yield forecasts were then used to calculate the area of each crop which would need to be grown as set out in Table 15.9.

TABLE 15.9 Family food needs, yields and areas

	Food needs (kg)	Yield per hectare (kg)	Hectares required
Maize	907	1344	0·67
Rice	454	4480	0·10
Sweet potatoes	907	6720	0·13
Cassava	454	5600	0·08
Groundnuts	91	672	0·13

The monthly labour requirements for these crops were then calculated and subtracted from the 48 man-days per month of family labour available. The residual family labour was available for the cultivation of cash crops. These results are given in Table 15.10.

TABLE 15.10 Labour allocation to food crops and surplus available

Crop	Hectares	J	F	M	A	M	J	J	A	S	O	N	D	Total
Maize	0·67	25	11	7	—	6	—	—	—	—	—	14	—	63
Rice	0·10	9	3	—	4	4	—	—	—	—	—	1	5	26
Sweet potatoes	0·13	—	9	3	2	—	—	—	—	—	—	—	—	14
Cassava	0·08	—	2	4	2	1	—	—	—	—	—	—	—	9
Groundnuts	0·13	2	—	—	—	6	6	—	—	—	—	3	7	24
Total	1·11	36	25	14	8	17	6	—	—	—	—	18	12	136
Surplus (48 minus total)		12	23	34	40	31	42	48	48	48	48	30	36	440

The gross margin method of partial budgeting was then used to determine which of the alternative cash crops would yield the highest gross margin subject to the constraints of the limited supplies of labour given in the last row of Table 15.10. In this case it was found that groundnuts (already included as a food crop) now considered as a cash crop were limited to 0·65 hectares by the labour constraint in May and December. The expected gross margin was £30 per hectare or £19·50 from 0·65 hectares. Cotton was limited to 1·01 hectares by the labour supply in December and January. At an expected gross margin of £50 per hectare this would yield £50·50. Cotton was therefore selected and expanded up to its limit of 1·01 hectares. The final farm plan for the initial season, therefore, consisted of 1·11 hectares of food crops and 1·01 hectares of cotton.

For the next season's plan, the input–output data were modified in the light of the first year's experience. In particular it was found that food requirements had been overestimated in the first year, and the food crop yields had turned out higher than expected. In the second year therefore it was planned to reduce the total food crop area to 0·7 hectares from the 1·11 proposed in the first year. This released additional family labour in every month. Furthermore, the farmer made a cash surplus in the first year and it was proposed to use approximately half this cash surplus to hire casual labour in the second year. In fact it was estimated that 113 mandays could be hired. Thus there was considerably more labour available for cash crops in the second year, both because less family labour was needed for food crops and because extra labour could be hired.

However, it was decided that the labour available for cash crops could not be used to grow cotton alone, because for technical reasons it was unsafe to grow cotton on the same land for more than two years in succession. This introduced a rotational constraint. The rotatable food crop area allowed 0·95 hectares of cotton, but this still left surplus labour resources. Hence a second cash crop was needed. After cotton, groundnuts were found to give the highest gross margin per man-day of labour in the critical months. In fact, for casual hired labour, available at any time of the year, groundnuts were expected to yield a higher gross margin per man-day than cotton. It was therefore decided to grow a combination of the two crops, cotton and groundnuts, in the ratio 2:1 in the rotation. This decision was made on subjective grounds rather than on the basis of some precise calculation. This combination of cotton and groundnuts in the area ratio of 2:1 was then expanded until all the available casual labour was accounted for. This resulted in a total of 2·23 hectares of cotton and 0·73 hectares of groundnuts.

Further modifications were introduced into the plan for the third year, but the approach was basically similar, in that:

(1) the necessary areas of food crops were estimated on the basis of dietary requirements and food crop yield estimates;

(2) the labour requirements of these crops were estimated month by month and deducted from the estimated family labour supply;

(3) a main cash crop was selected on the basis of gross margin per unit of limiting monthly labour resource, *but* the area was limited by a rotational constraint;

(4) a second cash crop was selected for rotating with the main crop again on the basis of the return to the most limiting resource;

(5) the ratio between the two crops was decided on a subjective basis;

(6) the combination of crops was expanded until one or other of the labour constraints was entirely accounted for.

GENERAL COMMENTS ON PROGRAMME PLANNING

Of course, programme planning need not follow precisely the same steps. In many cases it will not be thought necessary for the family to be entirely self-sufficient in food. Food crops could be treated like cash crops and only brought into the plan if it is profitable to do so. The first step of the programme then might be to select the enterprise which yielded the highest gross margin per unit of most limiting resource. Resources other than labour and working capital might be included in the list of possible constraints. In some regions land area or at least land of a particular type may be an important constraint. Other constraints may be due to technical or social requirements. Nevertheless, the general approach of satisfying the most limiting constraint in the way which yields the highest gross margin is common to most methods of programme planning.

However, we have already seen the difficulties which may arise in identifying the most limiting constraint. Furthermore, where there are several limiting constraints, the most profitable plan is likely to include a combination of the enterprises, rather than one single enterprise. This was the case in the example just described, the final plan including food crops, cotton and groundnuts satisfying at least three constraints, namely, the subsistence one, the peak labour one and the rotational constraint.

The general problem may be illustrated with a highly simplified imaginary example, with only two constraints, June labour and December labour, both restricted to 30 man-days, and two alternative enterprises, cotton and groundnuts. It is assumed that cotton will yield a gross margin per hectare of £50 and requires 20 man-days per hectare in June and 25 man-days per hectare in December. Groundnuts are assumed to yield a gross margin of £40 per hectare and to need 24 man-days per hectare in June and only 15 man-days per hectare in December. These figures set out in Table 15.11 have been so chosen that the peak labour need, and hence the critical constraint, differs for the two crops. It occurs in December for cotton and in June for groundnuts.

TABLE 15.11 Data for programme planning (hypothetical)

	Labour available	Cotton	Groundnuts
Gross margin (£)		50	40
June labour (man-days)	30	20	24
December labour (man-days)	30	25	15

The December labour constraint limits the maximum possible area of cotton to 30/25 = 1·2 hectares which will yield a total gross margin of £50 × 1·2 = £60. The June labour constraint limits the maximum possible area of groundnuts to 30/24 = 1·25 hectares which will yield a total gross margin of £40 × 1·25 = £50. Hence if choosing between growing cotton alone or groundnuts alone the highest gross margin and hence profit would come from growing 1·2 hectares of cotton. This would use up all the December labour but would leave some surplus labour in June. Thus 1·2 hectares of cotton would only need 1·2 × 20 = 24 man-days in June.

The question then arises as to whether it would pay to use the surplus June labour in producing groundnuts. But groundnuts need labour in December too, so groundnuts could only be introduced by reducing the area of cotton. Clearly the situation is now quite complicated. It is difficult to see immediately what feasible combination of cotton and groundnuts will yield the highest gross margin. The answer may be arrived at by the technique of linear programming to be discussed in the next chapter. In fact the answer is to produce 0·9 hectares of cotton and 0·5 hectares of groundnuts. This will yield a total gross margin of £65 as against £60 from cotton alone or £50 from groundnuts alone. The combination of crops exactly accounts for all the available labour in both months.

Programme planning is often recommended instead of linear programming for farm planning because the latter is considered to be too inflexible and time consuming. Short cuts are taken in programme planning, usually by relying, to some extent, on subjective judgement. The cost of this simplification is a reduction in the total expected gross margin.

Thus in the simple example just discussed, the programme planner might not worry about using up the surplus June labour available when cotton is produced alone. He might be willing to accept an expected gross margin of £60 instead of the £65 possible with a combination of crops. For this possible cost of £5 he would save planning time and effort and simplify the farm system. Of course, in practice rotational constraints might prevent the continuous growing of cotton but we are ignoring such constraints in this example. Another programme planning approach, used on the Tanzanian unit farm, is to decide on a suitable ratio of the two crops on technical grounds rather than on the basis of maximizing total gross margin.

Again subjective judgement may be used in deciding which constraints are the most limiting. Farmers or agricultural extension workers may be able to tell the planner which resources limit total production. If labour is the limiting factor they may be able to say from personal experience which is the busiest time of the year.

Indeed, the whole approach taken in programme planning is to some extent a matter for subjective judgement and is likely to depend on local circumstances. Thus programme planning on European farms usually starts from the assumption that land is the limiting factor. The first choice of enterprise is then the one with the highest gross margin per hectare. This approach would be inappropriate in many parts of Africa.

Under all conditions difficulties arise in handling the constraint of working capital, even though there may only be a limited sum available. This is because enterprises can finance each other and affect the overall capital requirement by their patterns of costs and returns relative to one another. Unspent or overspent capital balances are carried forward to swell or detract from current capital requirements. Furthermore short-term credit may be used to reduce peak capital requirements. In practice this means that the total farm requirement for capital at any time is not simply the sum of the capital requirements for the component enterprises at that time, when these have been calculated in isolation. The farm capital requirement is unique, at any time, to the pattern and size of the enterprises present and the credit facilities available. Working capital is therefore usually left out of the list of possible constraints in programme planning. However, working capital requirements of a plan can always be assessed by preparing a capital profile.

SUGGESTIONS FOR FURTHER READING

CLARKE, G. B. (1962) *Programme planning—a simple method of determining high profit production plans on individual farms,* O.E.C.D. documentation in food and agriculture, no. 45.

COLLINSON, M. P. (1969) Experiences with a trial management farm in Tanzania, *East African Journal of Rural Development,* vol. 2, no. 2, p. 28.

McFARQUHAR, A. M. M. (1962) Research in farm planning methods in northern Europe, *Journal of Agricultural Economics,* vol. 15, no. 1, p. 78.

16 Linear programming

THE ASSUMPTIONS

Linear programming is a systematic, mathematical procedure for finding the optimum, or best possible, plan for a given set of conditions. To use this method the conditions must be presented in the following form:

(1) a limited choice of several activities;
(2) certain fixed constraints affecting the choice;
(3) straight line (linear) relationships.

These basic assumptions made in linear programming need further explanation.

The alternative activities may correspond with the alternative enterprises which could be produced on the farm being planned, but the word 'activity' as used in linear programming does not necessarily mean the same as the word 'enterprise'. Thus buying activities may be considered in drawing up the plan. Furthermore, several different activities might be associated with the same enterprise, where several alternative methods of production are possible. Yams grown in a mixture represent a different activity from yams grown alone; irrigated cotton represents a different activity from rainland cotton. The choice must be limited to a suitable number of activities for calculation. For the simple examples to be worked out in this chapter, the number of activities is limited to three. However, larger problems only differ in the amount of arithmetic involved; the nature of the arithmetic is the same. For this reason most real linear programming problems are solved on electronic computers which are costly to use but which can deal with thousands of alternative activities. It could be argued that the number of alternative activities open to farmers is so large that some arbitrary selection is needed in any case, but this is obviously true for any planning technique.

The fixed constraints restrict the combinations of activities which are feasible. A plan which violates any constraint is assumed not to be feasible, which means that the constraint is assumed to be rigidly fixed. Thus if June labour, limited to 32 man-days, is a constraint, then a plan requiring 33 man-days of June labour would be rejected as not feasible in linear programming, although in practice it might be possible to manage the extra work. The constraints may be physical quantities of productive resources, they may be technical requirements such as that cotton may not be grown more than two years in direct succession, or they may be conditions which the farmer insists on for personal reasons. Many constraints are open ended, which means that they need not be met exactly but are either maximum or minimum limits. For instance although the total requirement for June labour cannot exceed 32 man-days, it may be less than this; the requirement for June labour must be equal to or less than the quantity available.

The assumption of linearity means that no matter how many units of a particular activity are included in the plan, the cost and return per unit remains the same. Thus if 1 hectare of maize needs 8 man-days of June labour and yields a gross margin of £10 then 4 hectares of maize are assumed to need $4 \times 8 = 32$ man-days of June labour and to yield a gross margin of $4 \times £10 = £40$. Likewise a quarter of a hectare is assumed to need $8/4 = 2$ man-days of June labour and to yield $£10/4 = £2·50$. The relationships between inputs and outputs are assumed to be straight lines. This can result in linear-programmed plans which include unrealistic fractions, particularly where livestock enterprises are involved. In a wide variety of problems the precision lost in rounding fractions to whole numbers is not sufficient to invalidate the solution.

The basic data necessary for linear programming consist of a specification of the alternative activities to be considered and the return or margin per unit of each activity. The constraints must also be listed and the total capacity of each estimated, together with the demands of each activity on each constraint. Where constraints consist of limited resource or input supplies, the demand of each activity on each of these constraints is known as an 'input–output coefficient'.

Linear programming then proceeds by a so-called iterative procedure, meaning a step-by-step, trial and error procedure to the optimum solution. The procedure is systematic in that the following features are involved.

1. There is a mechanical rule which determines, after each step, exactly what the next step is to be on the basis of the results of the trial just completed. One purpose of this feature is that it makes electronic computation possible. Computers do not generally

possess judgement of their own so they must be instructed precisely in the procedure to follow. In any event a mechanical rule stating what must be done at each succeeding trial in the trial and error procedure is useful because in a problem complicated by a great number of variables and inter-relationships, human judgement can go badly wrong and can result in an inefficient, even totally ineffective, search for the answer.

2. The mechanical rule is such that each step or iteration will yield a higher profit or gross margin than the previous one. This very important feature ensures that the calculations are working steadily closer to the desired result and not going off in a wrong direction.

AN EXAMPLE

We will take, as an example, a highly simplified case of a farm planning problem with just three alternative activities and four constraints. The objective is to maximize total gross margin. The constraints are land, limited to 4 hectares, labour in the peak-requirement months of June and December limited to 32 man-days and 30 man-days respectively, and total labour limited to 350 man-days. The three activities consist of three alternative crops, maize, groundnuts and cotton. Each hectare of maize requires 8 man-days of June labour, no labour in December and 80 man-days of labour in total and yields a gross margin of £10 per hectare. Each hectare of groundnuts requires 16 man-days in June, 20 man-days in December, 100 man-days in total and yields a gross margin of £40 per hectare. Each hectare of cotton requires 24 man-days in June, 25 in December, 150 in total and yields a gross margin of £50 per hectare. These data are tabulated in Table 16.1.

TABLE 16.1 Basic data for linear programming problem

| | Constraint level | Activities | | |
		Maize	Groundnuts	Cotton
Gross margin		£10	£40	£50
Constraint				
Land (hectares)	4	1	1	1
June labour (man-days)	32	8	16	24
December labour (man-days)	30	0	20	25
Total labour (man-days)	350	80	100	150

If we use the terms X_1, X_2, X_3 to represent the number of units (in this case hectares) of maize, groundnuts and cotton appearing in a plan and the term Z to represent the total gross margin we can express our problem as a set of algebraic relationships as follows. Note that the symbol $\leqslant$ means 'is less than or equal to' and $\geqslant$ 'is greater than or equal to'.

Maximize total gross margin Z where

$$Z = 10X_1 + 40X_2 + 50X_3 \qquad (1.1)$$

subject to the constraints

$$1X_1 + \quad 1X_2 + \quad 1X_3 \leqslant \quad 4$$
$$\text{(hectares of land}^1),$$
$$8X_1 + \quad 16X_2 + 24X_3 \leqslant \quad 32$$
$$\text{(June labour, man-days),}$$
$$0X_1 + \quad 20X_2 + \quad 25X_3 \leqslant \quad 30 \qquad (1.2)$$
$$\text{(December labour, man-days)}$$
$$80X_1 + 100X_2 + 150X_3 \leqslant 350$$
$$\text{(total labour, man-days),}$$

and the non-negativity requirements

$$X_1 \geqslant 0, X_2 \geqslant 0, X_3 \geqslant 0. \qquad (1.3)$$

$$(1)$$

This is the standard form for a linear programming problem. It consists of three parts, (1.1) the function (e.g. profit or total gross margin) whose value is to be maximized, which is called 'the objective function', (1.2) the ordinary structural constraints and (1.3) the non-negativity conditions on the variables.

The constraints are all open ended, thus the total area under the three crops can be equal to, or less than the total available area of 4 hectares. The computation of the optimal plan will be easier if we convert the inequalities to equalities or equations. This may be achieved by introducing new variables to represent unused land and labour. For instance if we let X_4 equal the number of hectares of unused land then we can write the land constraint equation as follows:

$$1X_1 + 1X_2 + 1X_3 + 1X_4 = 4.$$

This variable X_4 is known as a 'slack variable' or a 'disposal activity' since it allows for the non-use or disposal of land. A disposal activity differs from a real activity in that there are assumed

---

[1] Units are given here but they are immaterial to the solution of the set of equations.

to be no costs of not using land and of course no profits and the only resource used up is land itself. It requires no labour or other resource to leave land unused. Similarly we can introduce another slack variable X_5 to allow for non-use of June labour, so that instead of writing $8X_1 + 16X_2 + 24X_3 \leqslant 32$ we can write:

$$8X_1 + 16X_2 + 24X_3 + 1X_5 = 32.$$

There is obviously no input of land or even labour at other times of the year required for the disposal activity of not using labour in June. Slack variable X_6 makes provision for the non-use of December labour and X_7 for total labour. With the aid of these slack variables we can re-write our problem (1) in the so-called equality form:

Maximize $Z = 10X_1 + 40X_2 + 50X_3$

subject to the constraints

$$\left. \begin{array}{rrrrr} 1X_1 + & 1X_2 + & 1X_3 + 1X_4 = & 4, \\ 8X_1 + & 16X_2 + & 24X_3 + 1X_5 = & 32, \\ 0X_1 + & 20X_2 + & 25X_3 + 1X_6 = & 30, \\ 80X_1 + & 100X_2 + & 150X_3 + 1X_7 = & 350, \end{array} \right\} (2)$$

and the non-negativity requirements

$$X_1 \geqslant 0, X_2 \geqslant 0, X_3 \geqslant 0, X_4 \geqslant 0, X_5 \geqslant 0, X_6 \geqslant 0, X_7 \geqslant 0.$$

By adding four new non-negative slack variables, we have been able to change all the inequalities into equations except of course for the non-negativity requirements. These are included to ensure that none of the activities can occur at a negative level. This rather silly-sounding restriction is important partly because things like this are never obvious to an electronic computer, and without such a restriction, some activities might be reduced to a negative level. Even when solving the problem by hand, graphically or with a desk calculator, it is convenient to work according to a standard mathematical routine rather than to decide subjectively whether a plan is realistic. Hence it is desirable that the programming technique should ensure that these non-negativity constraints are fulfilled.

Having set out the example in the standard linear programming form we will solve it in three different ways. First we will use an intuitive approach to arrive at the optimum solution, secondly the problem will be solved graphically and finally a formal mathematical procedure will be described.

AN INTUITIVE APPROACH

The reader should be warned that this approach is somewhat tedious and can only be used for the very simplest of problems. However, if it is followed through, it will illustrate the reasoning behind the formal mathematical procedure.

Starting with the data set out in Table 16.1 we can see immediately that one feasible plan would be not to produce anything but to leave all the resources unused in disposal as set out in Table 16.2.

TABLE 16.2 Initial state of linear programming problem

Real activities in plan	None
Gross margin	£0
Available land (hectares)	4
Available June labour (man-days)	32
Available December labour (man-days)	30
Available total labour (man-days)	350

Clearly this plan is not very attractive since the total gross margin is zero but it does give a basic solution from which to move step-by-step towards an optimum.

Now for the first step we will choose the activity yielding the highest gross margin per unit and expand this as far as possible. In this case the chosen activity is cotton with a gross margin per unit of £50. To discover how far this activity can be expanded we must divide each of the constraint levels by the relevant requirement for this crop. Since there are 4 hectares of land available and each unit of cotton requires 1 hectare, the land constraint permits a maximum of 4 units of cotton. Each unit of cotton needs 24 man-days of June labour out of a total available supply of 32 so this constraint will only permit $32/24 = 1.3$ units to be produced. December labour limits the amount of cotton to $30/25 = 1.2$ units and the total labour supply restricts cotton to $350/150 = 2.3$ units. For cotton then, December labour is the most limiting constraint since this restricts the amount which may be produced to 1.2 hectares. Any larger area of cotton would not be feasible.

The introduction of 1.2 hectares of cotton will reduce the level of each of the other constraints accordingly. Thus the land disposal activity is reduced, from 4 hectares to 2.8. Similarly June labour in disposal is reduced by 1.2 times the requirement per hectare of 24 man-days, that is by 28.8 man-days, to leave a remainder of 3.2 man-days. The situation at the end of the first step or iteration is set out in Table 16.3.

We now need to investigate whether it is possible to improve upon this plan. It is apparent that the critical constraint at this stage is December labour since it is all in use; there is none in disposal. This suggests that we should select from the remaining two enterprises of maize and groundnuts the one which yields the highest increase in gross margin (return) per unit of this limiting resource. If groundnuts are introduced, they must compete with cotton for the available labour in December. Since the introduction of a unit of groundnuts requires 20 man-days of December labour, this means that 20 less must be used on cotton. Furthermore, since cotton production involves 25 man-days of December

TABLE 16.3 State at end of first iteration

Plan	1·2 hectares cotton
Gross margin	£0 + (1·2 × £50) = £60
Available land (hectares)	4 − (1·2 × 1) = 2·8
Available June labour (man-days)	32 − (1·2 × 24) = 3·2
Available December labour (man-days)	30 − (1·2 × 25) = 0
Available total labour (man-days)	350 − (1·2 × 150) = 170

labour per unit, a reduction of 20 man-days would cause a reduction of $20/25 = 0.8$ units. This means that 1 hectare of groundnuts substitutes for 0·8 hectares of cotton; the rate of technical substitution is 0·8. Now the return (gross margin) from a hectare of groundnuts is £40 and the opportunity cost is the return from 0·8 hectares of cotton which amounts to 0·8 × £50 = £40. The marginal net gain from transferring December labour from cotton to groundnut production is £40 − £40 = £0.

Thus there is nothing to be gained, at this stage, from transferring December labour from cotton production to groundnut production. However maize does not compete for labour in this critical period. It is a supplementary enterprise with respect to December labour so the rate of technical substitution and the opportunity cost of expanding maize are both zero. Maize should therefore be introduced into the plan and expanded as far as possible.

As before we discover how far this activity can be expanded by dividing each of the constraint levels by the relevant requirement for this crop. However the constraint levels have already been reduced as shown in Table 16.3. The limits set by each constraint on maize production are as follows:

Land	$2\cdot8/1 = 2\cdot8$
June labour	$3\cdot2/8 = 0\cdot4$
December labour	no restriction
Total labour	$170/80 = 2\cdot125$

Hence the critical constraint is now June labour and we can only introduce 0·4 hectares of maize before this constraint is exhausted. The situation at the end of the second iteration is given in Table 16.4.

TABLE 16.4 State at end of second iteration

Plan	1·2 hectares cotton + 0·4 hectares maize
Gross margin	$£60 + (0\cdot4 \times £10) = £64$
Available land (hectares)	$2\cdot8 - (0\cdot4 \times 1) = 2\cdot4$
Available June labour (man-days)	$3\cdot2 - (0\cdot4 \times 8) = 0$
Available December labour (man-days)	$0 - (0\cdot4 \times 0) = 0$
Available total labour (man-days)	$170 - (0\cdot4 \times 80) = 138$

Again we check whether it is possible to improve upon this plan. June labour is now the critical constraint restricting further expansion of the present plan. We therefore check to see which activity will yield the highest return to June labour by dividing the gross margin per unit by the June labour requirement per unit. This gives a gross margin per man-day of June labour of £1·25 for maize, £2·5 for groundnuts and £2·1 for cotton. Hence groundnuts yield the highest return to June labour so total gross margin can be increased by introducing groundnuts into the plan. What would be the effect of introducing 1 unit of groundnuts? The June labour required for groundnuts could be drawn from either maize or cotton, preferably the former since the return to June labour is lowest for maize. However, December labour *must* be drawn from cotton production since maize does not use any labour in this month. Now the rate of technical substitution of groundnuts for cotton in terms of December labour is 0·8 as we have seen, *but* at this stage of the plan, the transfer of December labour from cotton to groundnuts will release June labour, which in turn will allow the maize area to be increased. Thus if 1 unit of groundnuts requiring 16 man-days of June labour is introduced at the expense of 0·8 units of cotton requiring $0\cdot8 \times 24 = 19\cdot2$ man-days the June labour requirement is reduced by $19\cdot2 - 16 = 3\cdot2$ man-days.

This means that each unit of groundnuts introduced releases 3·2 man-days of labour in June. Now, although maize gives the poorest return to June labour, it does not require any December labour so it can be expanded to use up the extra June labour without competing with either of the other two activities. Each 3·2 man-days of June labour will allow $3·2/8 = 0·4$ units of maize to be introduced. Hence the total result of introducing 1 unit of groundnuts is a reduction of 0·8 units of cotton and an increase of 0·4 units of maize. The benefit is therefore £40 for the groundnuts plus $0·4 \times £10 = £4$ for the maize to set against an opportunity cost of $0·8 \times £50 = £40$. The marginal net gain from transferring December labour from cotton to groundnut production is $£44 - £40 = £4$. It therefore pays to make this change.

As before we discover how far this activity can be expanded by dividing each of the constraint levels by the relevant requirement for this crop. However, the area of cotton is now a constraint on the amount of groundnuts which can be introduced, since the groundnuts are substituting for cotton. Since there are 1·2 hectares of cotton in the plan, and each hectare of groundnuts substitutes for 0·8 hectares of cotton, the total area of groundnuts which can be introduced is $1·2/0·8 = 1·5$ hectares. This would permit an expansion of maize by $1·5 \times 0·4 = 0·6$ hectares. The other two constraints of land and total labour will permit this change as there is still a surplus of each of these, as set out in Table 16.5. Cotton is now eliminated from the plan.

TABLE 16.5 State at end of third iteration

Plan	1·5 hectares groundnuts and $0·4 + 0·6 = 1$ hectare maize (1·2 hectares cotton go out)
Gross margin	$64 - (1·2 \times 50) \ + (1·5 \times 40) \ + (0·6 \times 10) = \ 70$
Available land (hectares)	$2·4 + (1·2 \times 1) \ - (1·5 \times 1) \ - (0·6 \times 1) = \ 1·5$
Available June labour (man-days)	$0 + (1·2 \times 24) \ - (1·5 \times 16) \ - (0·6 \times 8) \ = \ 0$
Available December labour (man-days)	$0 + (1·2 \times 25) \ - (1·5 \times 20) \ - (0·6 \times 0) \ = \ 0$
Available total labour (man-days)	$138 + (1·2 \times 150) - (1·5 \times 100) - (0·6 \times 80) = 120$

Once more we must check whether it is possible to improve upon this plan. Firstly if cotton is re-introduced each unit will compete with groundnuts for December labour and with both groundnuts and maize for June labour. The introduction of 1 unit of cotton would require a reduction of $25/20 = 1\cdot25$ units of groundnuts to release the necessary December labour. This would also release $1\cdot25 \times 16 = 20$ man-days of June labour but 1 unit of cotton requires 24 man-days in this month, that is 4 extra. The maize activity must therefore be reduced by $4/8 = 0\cdot5$ units to release this June labour. The total opportunity cost of introducing 1 hectare of cotton is therefore $1\cdot25 \times £40 + 0\cdot5 \times £10 = £55$. Since the gross margin from 1 hectare of cotton is only £50 this cannot be profitable.

Secondly, we must consider the possibility of expanding groundnut production. In fact this is impossible since all the existing supply of December labour is required for the present area of groundnuts.

Finally, what is the effect of expanding maize production? Obviously it would compete with groundnuts for labour in June. One unit of maize requires 8 man-days in this month. It would therefore substitute for $8/16 = 0\cdot5$ units of groundnuts. The opportunity cost is therefore $0\cdot5 \times £40 = £20$ for a benefit from 1 hectare of maize of only £10. Hence we can say that it is in no way possible to increase the total gross margin and hence improve upon the plan given in Table 16.5. This now is the optimum solution.

Additional information may be derived on the effect of relaxing the constraints individually. Thus, we can calculate the effect on total gross margin of increasing each of the resource constraints in this example by 1 unit. Firstly, if the land constraint is increased by 1 hectare, the total gross margin would be unaffected. This is because there is already a surplus in the land disposal activity. It is impossible to increase the gross margin unless labour resources are increased with the given set of activities. Similarly, there is nothing to be gained from increasing total labour supply if the labour supply in the months of June and December remains constant. This again occurs because there is surplus labour in total.

However, there is no surplus labour in June, so this is a limiting constraint on production. An extra man-day of labour in this month would enable the maize activity to be expanded. It would not enable the area of groundnuts to be increased because this is limited by December labour. Since 1 hectare of maize requires 8 man-days of June labour, 1 extra man-day would allow an expansion of one-eighth of a hectare of maize. This would yield a gross margin of £10/8 = £1·25. The gain from an increase of 1 man-day of July labour is therefore £1·25.

An extra man-day of December labour would allow the area of groundnuts to be increased by one-twentieth of a hectare. This would yield a gross margin of £40/20 = £2. However, groundnuts compete with maize for June labour, so this expansion could only be achieved at the cost of reducing the maize area. In fact one-twentieth of a hectare of groundnuts requires 16/20 = 0·8 man-days of June labour, which would be provided by reducing the maize area by 0·8/8 = 0·1 hectares. The cost of this is 0·1 × £10 = £1. Thus the net gain from 1 man-day of December labour is £2 − £1 = £1.

These estimates of the net gain to be obtained by increasing each resource constraint by 1 unit, all other constraints remaining fixed, are, of course, estimates of the marginal value products of these resources. If all the constraints are increased together, then the whole optimum plan can be expanded to scale. The basic assumptions of linear programming imply that there are constant returns to scale.

A GRAPHICAL APPROACH

Another approach, which may be easier to follow, is a graphical one. However, it is only suited to problems where there are two alternative activities, so again it can only be used for relatively simple problems. Knowing, as we do, that cotton does not appear in the final solution of our problem set out in Table 16.1, we can reduce this to a two-activity problem and solve it graphically. Since we are solving exactly the same problem in a different way we will hope to arrive at the same solution.

Basically we will use the production possibility curve approach of Chapter 4, that is, we will plot possible combinations of groundnuts and maize on a graph. Consider first the land constraint which may now be written: $1X_1 + 1X_2 \leqslant 4$. This constraint would be satisfied without leaving any land in disposal by producing 4 hectares of maize ($X_1 = 4$) or 4 hectares of groundnuts ($X_2 = 4$) or any intermediate combination such as 2 hectares of maize and 2 of groundnuts. This is shown on a graph in Figure 16.1 by the line AB. Any point on this line represents full use of the land resource constraint and clearly satisfies the inequality since it fits the equation $1X_1 + 1X_2 = 4$. In addition, any point such as C

which lies below and to the left of this line also satisfies the inequality since it involves values of X_1 and X_2 smaller than those which use up all the land, that is point C implies some land unused. Thus the area OAB represents all feasible combinations of maize and groundnuts subject to the single land constraint. Any point outside this area violates either the land constraint (point D) or the non-negativity constraints (points E and F).

We will now add the constraint line for June labour. If all this resource is used in maize production, $32/8 = 4$ units would be

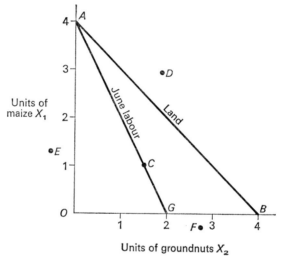

FIGURE 16.1 Land and June labour constraints. The land constraint limits feasible combinations of maize and groundnuts to points within the triangle ABO; the June labour constraint to points within the triangle AGO.

feasible, whereas if entirely devoted to groundnut production 32 $16 = 2$ units would be feasible. Thus the line AG in Figure 16.1 represents the line $8X_1 + 16X_2 = 32$ and encloses all possible combinations of maize and groundnuts which satisfy the June labour constraint. The constraints for December labour and total labour are represented in a similar fashion in Figure 16.2. Since maize does not require labour in December, the line for this constraint rises vertically from the maximum possible level of groundnut production.

It should now be clear that in Figure 16.2 the shaded area represents all feasible combinations of maize and groundnuts, and the line ACH is the production possibility boundary. In this case it is

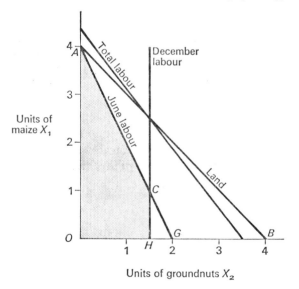

FIGURE 16.2 The feasible area. Feasible solutions are restricted to the shaded area *OACH*. Note similarity to Figure 4.1.

not a curve but a series of linear segments, but as in Figure 4.1 (page 73) it is concave to the origin. Between H and C, where the boundary rises vertically, maize is supplementary to groundnuts in the use of December labour. It should also be noted in this case that total labour is not a limiting resource over any part of the boundary. No matter what combination of maize and groundnuts is found to be optimal there is bound to be some surplus labour in total; the labour disposal activity must have a positive value. The same is true for land except for the single possibility at point A where it would all be used in maize production.

The maximum gross margin combination is now found by adding iso-revenue lines as was done in Figure 4.2. Iso-revenue lines are shown superimposed on the production possibility boundary from Figure 16.2, in Figure 16.3. Each of these lines joins all the combinations of units of maize (X_1) and units of groundnuts (X_2) which will yield a given total gross margin $(\mathcal{Z})$. Since the gross margin per unit of maize is £10 and the gross margin per unit of groundnuts is £40, total gross margin is calculated by the equation $\mathcal{Z} = 10X_1 + 40X_2$. Now if we set $\mathcal{Z}$ equal to 10 for the £10 iso-revenue line, we have a set of solutions ranging from $X_1 = 1$ and $X_2 = 0$ to $X_1 = 0$ and $X_2 = 0.25$. Higher iso-revenue lines are calculated in the same way.

It is now clear that the maximum total gross margin that can be obtained is £70 at point *C*. This is the highest 'rung' that is feasible on the 'step ladder' of increasing incomes. Incomes higher than that obtained at point *C* are not feasible. This point represents a combination of 1·5 units of groundnuts and 1 unit of maize, which is identical with the optimum solution previously obtained. Reference back to Figure 16.2 will further confirm the other results obtained earlier, namely that June labour and December labour are the only two limiting constraints. Some land and some of the total labour supply is unused in the final plan.

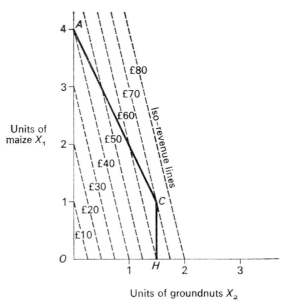

FIGURE 16.3 Iso-revenue lines and the optimum solution. The optimum solution is found at point *C*. Note similarity to Figure 4.2.

Since a move to a higher iso-revenue line always increases profits, it follows that the optimum solution of a linear programming problem will always be on the boundary of the feasible region. Whatever the relative gross margins of maize and groundnuts are, an optimum solution would always be found at one of the four corners $(0, A, C$ and $H)$ of this region.

This important result represents the basic theorem of linear programming, namely that in seeking the optimum solution we need only consider the results for the corner points of the feasible region. In this way linear programming greatly simplifies the problem of finding the optimum. Even in this very simple example with only two alternative activities of maize growing and ground-nut growing, the number of feasible combinations is infinite. Any combinations of the two, represented by a point inside the area *OACH*, would be feasible. If planning by simple budgeting we could estimate the total gross margin from even hundreds of these combinations and still never find the optimum, *but* the linear pro-gramming theorem tells us we need only consider the corners, of which in this case there are only four. The iterations involved in the formal solution of a linear programming problem represent movements from one corner to another more profitable corner of the feasible region.

This principle still holds even when we have many activities and constraints so that the graphical method cannot be used. At each corner we are in effect reducing certain of the activities (both slack and ordinary) to zero so that we are left with a solution where the number of non-zero-valued activities (both slack and ordinary) is *exactly equal* to the number of constraints in the problem. Such a solution is known as a 'basic solution'. In Figure 16.2 point *A* is an unusual exception since both land and June labour are used up exactly but at each other corner *O*, *H* and *C* the number of non-zero activities is equal to the number of constraints, in this case four. For instance point *H* has the four non-zero activities of 1·5 units of groundnuts, 8 units of June labour in disposal, 200 units of total labour in disposal and 2·5 units of land in disposal. Thus the iterations involved in the formal solution of a linear programming problem represents movements from one basic solution to another more profitable basic solution. This has the important practical result that the number of activities in the final solution will always equal the number of constraints. Thus if the farmer has say seventy alternative activities but only three constraints the optimum solu-tion will only include three of the original seventy activities.

A FORMAL MATHEMATICAL APPROACH

It is now proposed to describe the solution of our linear program-ming problem by a series of mechanical, mathematical steps. By following these steps it should be possible to solve slightly more complicated problems than our present example without having to think about the logic at every stage. Furthermore this gives a rough picture of the calculations carried out by an electronic com-puter when it is used for linear programming.

The method described here in outline is the simplex method which is the one most frequently used to deal with problems such as our example and with other problems which are vastly larger.[2] The reader may relate the various steps of the calculation to those described in the intuitive approach.

To start with, the initial set of equations defining our problem (2) are rearranged with the slack variables on the left as in (2') below:

Maximize $Z = 0 + 10X_1 + 40X_2 + 50X_3$

subject to

$$
\begin{aligned}
X_4 &= 4 - 1X_1 - 1X_2 - 1X_3, \\
X_5 &= 32 - 8X_1 - 16X_2 - 24X_3, \\
X_6 &= 30 - 0X_1 - 20X_2 - 25X_3, \\
X_7 &= 350 - 80X_1 - 100X_2 - 150X_3, \\
& \quad X_1 \text{ to } X_7 \geqslant 0.
\end{aligned}
\qquad (2')
$$

This in turn may be set out in a table or matrix of coefficients (Table 16.6) corresponding to the first basic solution with the real activities at zero level and the slack variables at a maximum. In Table 16.6 the columns represent real activities which are at zero level initially and all the rows represent disposal activities which are positive and at a maximum initially. The column marked R may be ignored for the moment.

TABLE 16.6 Linear programming matrix

		X_1 Maize	X_2 Groundnuts	X_3 Cotton	R
Z	0	10	40	50	
X_4 Land disposal	4	−1	−1	−1	4
X_5 June labour disposal	32	−8	−16	−24	1·33
X_6 December labour disposal	30	0	−20	**−25**	1·2
X_7 Total labour disposal	350	−80	−100	−150	2·3

Essentially each step or iteration now consists in substituting one of the columns for one of the rows. Thus in the first iteration we substitute a real variable for a slack variable. The element which occurs where the selected column and the selected row intersect is known as the *pivot* or pivot element. That column may then be referred to as the pivot column and the row as the pivot row.

[2] This description of the simplex method differs slightly from that given in some other texts (such as Heady, E. O., and Candler, W. (1958) *Linear programming methods*, Iowa State University Press) *but* the principles are identical. This method is described in Baumol, W. J. (1972) *Economic theory and operations analysis*, 3rd ed., Englewood Cliffs, Prentice-Hall.

Columns always represent zero valued variables whereas rows represent positive valued variables. Thus substitution of a column for a row means that the activity represented by the pivot column is being introduced to a positive level at the expense of reducing the pivot row activity to zero. The non-negativity constraints require that the levels given in the first column of the matrix must never be negative. We can now describe the calculations involved at each iteration by a series of simple rules.

Rule 1. Choice of pivot column. The column with the highest positive top element (the highest increase in gross margin per unit) is chosen as pivot column. In Table 16.6 the pivot column is that for cotton X_3 since 50 is the highest top element. Once there are no positive top elements the total profit cannot be increased and the optimal solution has been found.

Rule 2. Choice of pivot element. Take each negative element in the pivot column and divide it into the corresponding element in the first column. The element for which the resulting quotient is smallest in absolute value (i.e. smallest, ignoring sign) must be chosen as pivot element. This tells us which constraint is the most limiting.

For our example in Table 16.6 these ratios have been added in the column marked R. The smallest of these in absolute value is 1·2 which occurs in the December labour disposal X_6 row. Cotton will therefore be introduced at the expense of reducing December labour disposal to zero. The pivot element, where the cotton column intersects with the December labour disposal row with a value of -25, is printed in bold type in Table 16.6. We now make the necessary calculations to complete the first iteration and a new matrix.

Rule 3. New pivot element. The element which replaces the old pivot (-25) is simply the number 1 divided by the old pivot element $(1/-25 = -0·04)$.

Rule 4. Other pivot row elements. Any other element in this row of the matrix is obtained by changing the sign of the corresponding old element and dividing by the pivot element. Thus in our example, the old December labour disposal row which now becomes the cotton row in the new matrix is calculated as $-30/-25 = 1·2$, $0/-25 = 0$ and $+20/-25 = -0·8$, together with the last element replacing the old pivot of $-0·04$.

Rule 5. Other pivot column elements. Any other element in the column which contained the old pivot is replaced by the corresponding old element divided by the old pivot. In our example, the old cotton column which now becomes the new December labour column is calculated as follows: $50/-25 = -2$, $-1/-25 = +0.04$, $-24/-25 = +0.96$, pivot element, and $-150/-25 = +6$.

Rule 6. All other elements. This is best illustrated by means of an example. Thus we will consider the element in the top left-hand corner. In the old matrix this is a zero. Consider its position in relation to the pivot element and find the corner elements, one in the pivot row and one in the pivot column which, together with the pivot element and the element to be replaced, form the four corners of a rectangle. Table 16.7 is a reproduction of Table 16.6

TABLE 16.7 Elements involved in calculating new corner element

		X_1	X_2	X_3
z	0	10	40	50
X_4	4	-1	-1	-1
X_5	32	-8	-16	-24
X_6	30	0	-20	**-25**
X_7	350	-80	-100	-150

with the old element to be replaced underlined and the two other corner elements printed in italic type.

Now use the formula:

New element = corresponding old element
$$- \frac{\text{product of the two (other) corner elements}}{\text{old pivot element}}.$$

Thus the new term in the top left-hand corner becomes

$$0 - \frac{50 \times 30}{-25} = +60.$$

The remaining elements are calculated in the same way to yield a new matrix representing the result of the first iteration. This new matrix is completed in Table 16.8.

Now the highest positive top element is found in the X_1 maize column so this is the pivot column. The R column is calculated as before and the pivot row is found to be the X_5, June labour disposal row. Again a new matrix is completed for the second iteration. The whole process is repeated until an optimum solution is found. The remaining iterations are shown in Table 16.9.

TABLE 16.8 New matrix resulting from first iteration

		X_1	X_2	December labour X_6	R
Z	60	$+10$	0	-2	
X_4	2·8	-1	$-0·2$	$+0·04$	2·8
X_5	3·2	-8	$+3·2$	$+0·96$	0·4
X_3 cotton	1·2	0	$-0·8$	$-0·04$	∞
X_7	170	-80	$+20$	$+6$	2·125

The second matrix in Table 16.9 represents the optimum solution since the top figures in the three activity columns are all negative. The introduction of any of these three activities into the plan would reduce the total gross margin. The column labelled X_5 represents June labour disposal. The top figure in this column tells us that an increase of 1 man-day in June labour disposal would reduce the total gross margin by £1·25. Likewise the top figure in the final column tells us that another unit of December labour in disposal would reduce the total gross margin by £1. These figures therefore represent the marginal value product of the last available unit of labour in each of these two months. The top figure in the X_3 (cotton) column tells us that the introduction of 1 unit of cotton

TABLE 16.9 Remaining iterations of linear programming problem

Second iteration

		X_5	X_2	X_6	R
Z	64	$-1·25$	$+4$	$-0·8$	
X_4	2·4	$+0·125$	$-0·6$	$-0·08$	4
X_1	0·4	$-0·125$	$+0·4$	$+0·12$	
X_3	1·2	0	$-0·8$	$-0·04$	1·5
X_7	138	$+10$	-12	$-3·6$	11·5

Third iteration

		X_5	X_3	X_6
Z	70	$-1·25$	-5	-1
X_4	1·5	$+0·125$	$+0·75$	$-0·05$
X_1	1	$-0·125$	$-0·5$	$+0·10$
X_2	1·5	0	$-1·25$	$-0·05$
X_7	120	$+10$	$+15$	-3

into the plan would reduce the total gross margin by £5. It implies that if the gross margin per unit of cotton increased by £5, the gross margins for the other crops remaining constant, then cotton would come into the optimum plan.

The first column of the matrix gives details of the final plan. The top figure is the total gross margin and the remaining elements in this column show the level at which activities are included in the plan. Thus there are 1·5 units of land in disposal, 1 unit of maize, 1·5 units of groundnuts and 120 units of total labour in disposal. The remaining figures in the matrix show the rates at which activities excluded from the plan substitute for the combination of activities which are in the plan. For example the cotton column (X_3) tells us that 1 unit of cotton can be substituted for 0·5 units of maize X_1 and 1·25 units of groundnuts X_2 and at the same time would release 0·75 units of land X_4 and 15 units of total labour X_7. The result of this substitution, as we have already seen, would be a reduction of £5 in the total gross margin.

LINEAR PROGRAMMING AND THE THEORY OF PRODUCTION

At first sight it would appear that the assumption of linear relationships, which is fundamental to linear programming, is incompatible with the law of diminishing marginal returns. However, closer analysis of our simple example will show that in linear programming, marginal returns do in fact diminish. This is already implied by the concave shape of the production possibility boundary *but* it is brought out more clearly if we consider our example in terms of increases in the use of the variable input land, with a fixed supply of labour. The total product graph is shown in Figure 16.4.

The first 1·2 hectares of land are used in cotton production at a marginal value product per hectare of £50. Cotton production cannot be extended beyond 1·2 hectares because the area is limited by the supply of December labour. The next 0·4 hectares of land are used in maize production at a marginal value product of only £10 per hectare. Thus the marginal value product has diminished from £50 to only £10. Now that 1·6 hectares of land are in use, June labour limits further expansion. However, more land can be brought into production by substituting groundnuts and maize for cotton. For an expansion in land use from 1·6 to 2·5 hectares, that is an increase of 0·9 hectares, the increase in gross margin is only £6 (£70—£64). The marginal value product per hectare over this

0·9 hectares is, therefore, £6/0·9 = £6·67. It has again diminished. Beyond 2·5 hectares of land, further inputs yield a marginal value product of zero. The maximum total value product of £70 is therefore obtained when 2·5 hectares of land are in use.

Clearly, if it was possible to increase inputs of labour in December and June along with the land resource, cotton production could be expanded further yielding a constant marginal value product of £50. If, in this example, the total supply of labour could be expanded too, then all 4 hectares of land could be used in cotton production. Thus there are constant returns to scale. The marginal product of a single resource only diminishes as other resources become limiting.

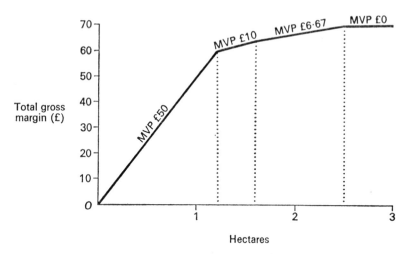

FIGURE 16.4 Linear programming and diminishing marginal returns. Marginal returns diminish but not along a smooth curve as in Figure 2.2.

However, decreasing returns to scale within a single enterprise can be written into the programme, if the enterprise is treated as several different activities depending upon its scale. To illustrate this, Figure 16.4 may now be considered as a graph relating total gross margin to number of hectares grown of a single enterprise, cotton. The first activity is cotton production up to 1·2 hectares yielding a gross margin per hectare of £50. There is an additional constraint which limits this activity to 1·2 units. The second activity is cotton production from 1·2 up to 1·6 hectares which yields only £10 per hectare. This also is limited by a constraint. Likewise the third activity, cotton production from 1·6 to 2·5 hectares, yields only £6·67 per hectare. It is also possible to allow for increasing returns to scale although a modified method of programming known as 'separable programming' is necessary.[3]

In many respects the assumption of constant marginal returns over a range of variable inputs is more realistic than the theoretical assumption of a smooth curve relating inputs and outputs. Furthermore, the complicated problem of allocating a fixed labour supply to enterprises with varying seasonal requirements is more readily handled by linear programming than by production function analysis. Working capital constraints can also be incorporated in a linear programme, although capital differs from labour in that if it is unused it accumulates. Thus a particular enterprise may either draw upon or contribute to the supply of capital in each month.[4]

However, the assumption of rigidly fixed resources is a little unrealistic in many cases. In practice if labour in December becomes a limiting constraint, the farmer will find some means of increasing the labour supply in that month. He and his family would undoubtedly be prepared to work harder and he might well find it profitable to hire labour. The linear programme may be made more flexible to meet this point in two ways. First, the possibility of hiring labour may be written in as an alternative buying activity, with a negative contribution to total revenue and a positive contribution to the labour constraint. As a second alternative, the linear programme may be repeated with the constraints set at different levels. This will give a measure of the effect on total profit of changing the resource mix. Repeated linear programming with changes in the resource mix or the unit revenues of the activities is known as 'parametric linear programming'. It is most useful for measuring the effect of chunky inputs such as regular labour or fixed capital. The economics of introducing tractors in both Kenya and Uganda have been investigated in this way.[5]

An alternative objective might be to minimize risk for a given minimum acceptable level of income, which is then a constraint. This and other approaches have been used to allow for risk within

[3] See Feldman, D. (1969) *Decreasing costs in tobacco production—an application of separable programming*, University of Sussex, Institute of Development Studies, Communications Series no. 40.

[4] See Stewart, J. D. (1961) Farm operating capital as a constraint, a problem in the application of linear programming, *The Farm Economist*, vol. 9, no. 10, p. 463.

[5] Clayton E. S. (1963) *Economic planning in peasant agriculture*, University of London, Wye College, Department of Agricultural Economics; Lea, J. D. and Joy, L. (1963) Modern arable farming in Uganda, *Empire Journal of Experimental Agriculture*, vol. 31, no. 122, p. 137.

the linear programming framework.[6] However, all these approaches require estimates of the risks or variation in inputs and returns for each activity and such information is not readily obtained.

OTHER APPLICATIONS

The technique of linear programming has many applications besides its use in farm planning just described. In fact it may be used to find an optimum or 'best' solution to *any* problem with the characteristics listed at the beginning of this chapter, namely:

(1) a limited choice of several alternative activities;
(2) certain fixed constraints affecting the choice;
(3) straight line (linear) relationships;

provided that the objective can be clearly defined and measured. It is therefore quite a flexible planning tool with widespread uses. Within this same framework many problems concerned with industrial production, the marketing of all types of commodities and regional development programmes can be investigated.[7]

Moreover, the technique is not limited to problems where the objective is to maximize profit or gain in some sense such as employment opportunities. It is equally useful where the objective is to minimize something, such as the cost of carrying out industrial processes or servicing biological needs. A typical example of the latter, drawn from the agricultural sector, is the use of linear programming to determine a minimum-cost feed mix for livestock, which will provide certain dietary requirements. For this problem the alternative activities consist of the various possible ingredients of the feed mix, such as maize, groundnut cake, sorghum, or fishmeal. The constraints are the energy, protein, fibre and bulk requirements of the livestock for which it is intended.

A simplified example involving four alternative ingredients and only two constraints is given below.

---

[6] See McFarquhar, A. M. M. (1961) Rational decision making and risk in farm planning, *Journal of Agricultural Economics*, vol. 14, no. 4, p. 552; McInearney, J. P. (1969) Linear programming and game theory models: some extensions, *Journal of Agricultural Economics*, vol. 20, no. 2, p. 267; also Boussard, J. M. and Petit, M. (1967) Representation of farmers' behaviour under uncertainty with a focus-loss constraint *American Journal of Agricultural Economics*, vol. 49, no. 4, p. 869.

[7] e.g. see Hall, M. (1970) *A linear programming approach to regional agricultural planning*, East African Agricultural Economics Society Conference (mimeograph).

Let X_1 = units of maize (kg) at 2p per kg,
$\quad X_2$ = units of sorghum (kg) at 1·5p per kg,
$\quad X_3$ = units of groundnut cake (kg) at 2·5p per kg,
$\quad X_4$ = units of fishmeal (kg) at 6p per kg.

Objective: Minimize cost of ration $\mathcal{Z}$

where $\mathcal{Z} = 2X_1 + 1\cdot5X_2 + 2\cdot5X_3 + 6X_4$

subject to the constraints

$0\cdot8X_1 + 0\cdot7X_2 + 0\cdot8X_3 + 0\cdot6X_4 \geqslant 2\cdot5$ (minimum energy TDN),

$0\cdot1X_1 + 0\cdot1X_2 + 0\cdot4X_3 + 0\cdot6X_4 \geqslant 0\cdot6$ (minimum protein),

and the non-negativity requirements

$X_1 \geqslant 0, \; X_2 \geqslant 0, \; X_3 \geqslant 0, \; X_4 \geqslant 0.$

This problem can be solved by linear programming provided a few modifications are made. First, slack variables must be introduced to convert the inequalities into equalities. In this case, however, they are minimum constraints which may be exceeded so the slack variables must be subtracted, not added as for maximum constraints. Hence if we arrive at a ration which provides 2·7 kg of TDN (total digestible nutrients) we must subtract 0·2 (the slack variable $X_5 = 0\cdot2$) from this number to arrive at the 2·5 kg requirement.

However, we cannot find an initial basic solution to this problem as we did before by setting all but the slack variables equal to zero. This is not now possible since it would make the slack variables negative. The minimum constraints cannot be met by setting the real activities, that is the alternative ingredients, at zero level. This problem is overcome by introducing imaginary additional ingredients which we may call Q_1 and Q_2 and which have the characteristics that 1 kg Q_1 contains 1 kg of TDN but no protein and 1 kg Q_2 contains 1 kg of protein but no energy. The price of these imaginary ingredients is then set very high. They may be thought of as very expensive pure extracts of the required energy or protein. We then start from an initial basic solution consisting of these imaginary ingredients which therefore is a very costly ration and then work step-by-step towards a less costly diet, consisting of real ingredients and slack variables.

It will be noted that since there are only two constraints in this example only two activities are included in the optimum solution.

In fact the least-cost ration which will provide 2·5 kg of TDN and 0·6 kg of protein contains 2·6 kg of sorghum and 0·85 kg of groundnut cake. The solution is illustrated graphically in Figure 16.5,

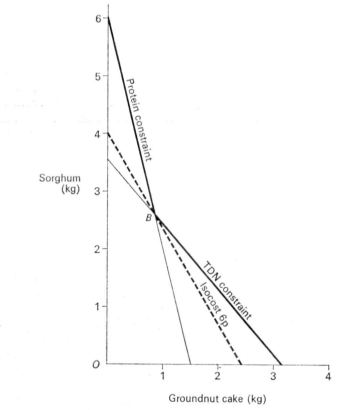

FIGURE 16.5 A least-cost linear programme. The feasible area lies above and to the right of the constraint lines. The least-cost combination is found at point *B*. Note similarity to Figure 2.9.

which should be compared with the isoquant graph of Figure 2.9 (page 38). We are in effect finding a least-cost combination of resources by linear programming in the present example.[8]

The least-cost combination of resources is naturally dependent on the relative prices of the resources. Changes in the relative

___

[8] This technique is described in more detail in Heady and Candler, *Linear programming methods*; Dent, J. B. and Casey, H. (1967) *Linear programming and animal nutrition*, London, Crosby Lockwood; and Upton, M. (1965) Linear programming rations for pigs, *Nigerian Journal of Economic and Social Studies*, vol. 7, no. 2, p. 221.

prices of ingredients alter the composition of the cheapest diet. Where livestock feeds are mixed on a large scale, linear programming of least-cost rations may be needed at regular intervals to allow for changes in the relative prices of ingredients.

LINEAR PROGRAMMING A KENYAN FARM

So far we have used simple, imaginary examples to show the method of linear programming. Since the method of solving the larger problems met with in real life is essentially the same, there is little to be gained in working out a solution to a more realistic example. However, there is a certain amount of skill involved in drawing up the initial linear programming matrix, a job which must be done whether the solution is to be found on an electronic computer or by calculation on paper. It is therefore useful to consider how one particular research worker put his matrix together.

The example chosen was an early application of linear programming to a real farm situation in the Nyeri district of Central province, Kenya.[9] A budget for this farm is given in Chapter 14 but it was shown that this plan could be improved upon by linear programming. Several programmes were prepared with slight variations in the constraints and activities. The programme discussed here for which the matrix is given in Table 16.10 is the second of the series, in which a rotational constraint is included and tea is considered as a possible activity. Note that if the problem is to be solved by the formal mathematical method described earlier, the signs would be changed on all the input–output coefficients. In fact every coefficient would have a minus sign except for that at the intersection of the 'ley and milk column' and the 'arable/ley ratio' row which would become $+4$. However, it is convenient to set out the data in its present form, which is suitable for many computer programmes.

All enterprises and land resources were measured in acres. These units are not converted to hectares here, as this might introduce errors of rounding. However, the units used have no effect on

[9] Clayton, *Journal of Agricultural Economics*, vol. 14, no. 3, p. 337. Other important applications in Africa not yet mentioned are McFarquhar, A. M. M., and Evans, A. (1957) Linear programming and the combination of enterprises in tropical agriculture, *Journal of Agricultural Economics*, vol. 12, no. 4, p. 474; Heyer, J. (1967) *The economics of small-scale farming in lowland Machakos*, Nairobi, University College, Institute for Development Studies, Occasional Paper no. 1; and Olayide, S. O., Olowude, S. O., and Oni, S. A. (1971) Application of linear programming to farm planning in developing countries: a case study of arable farming in Western Nigeria, *Association for the Advancement of Agricultural Sciences in Africa Journal*, vol. 1, no. 1.

TABLE 16.10 Linear programming matrix for case-study farm in Kenya (after Clayton)

	English potatoes		Maize		Beans		Ley and milk	Coffee and mulch	Tea
	LR	SR	LR	SR	LR	SR			
Z ($£$)	22·10	22·10	12·35	9·85	12·55	12·55	12·50	95·00	100·00
Monthly labour constraint[a]									
March 600	89	0	73	0	101	0	17	135	85
April 600	76	0	55	0	39	0	18	175	85
July 600	189	0	0	0	41	0	34	32	98
October 600	0	89	0	73	0	101	17	110	85
November 600	0	76	0	85	0	50	17	143	85
December 600	0	100	0	70	0	49	17	142	177
January 600	0	89	0	0	0	41	32	169	85
February 600	0	100	0	95	0	0	31	189	85
Land[b] LR 7·37	1	0	1	0	1	0	1	1	1
Land[b] SR 7·37	0	1	0	1	0	1	1	1	1
Arable/ley ratio 0	3	0	3	0	3	0	−4	0	0

Note. LR = long rains; SR = short rains.
[a]man-hours. [b]acres.

the optimum solution and its profitability. The important thing is to be consistent in using the same units to measure both the input–output coefficients and the constraints. The coefficients for each activity were obtained from secondary sources and it is not known how accurately they reflect the potential results on the individual farm considered.

In this programme, one livestock and five crop enterprises are considered. These embrace the range of ecological possibilities as far as arable crops are concerned except for wheat, for which labour data were not available, and peas which have labour requirements similar to beans, but a lower return per acre. Millet and cassava are omitted because the returns were too low to justify their inclusion.

Double cropping is normal practice in Kenya and the arable crops considered can be grown in both the 'long' and 'short' rains periods. As its name implies the short rains season is of shorter duration with a lower rainfall. Thus for each arable crop there are two alternative activities, a long rains crop or a short rains crop. It is assumed that the yield is the same in each season except for maize which yields less in the short rains season. This is reflected in a lower return per unit. Grass leys and milk production are treated together as a single activity. Stocking is assumed at the rate of 1 cow per acre, hence labour needs and the financial returns relate to 1 cow and 1 acre of ley. All herd replacements are assumed to be bought in. No alternative uses of grassland such as beef or sheep production are considered, nor is the possibility of varying the stocking rate. Coffee and mulch are also treated together as a single activity. On technical grounds it is claimed that a mulch is essential for high coffee yields, hence coffee requires an equal area of napier grass mulch. One unit of coffee and mulch therefore consists of $\frac{1}{2}$ acre of coffee and $\frac{1}{2}$ acre of mulch. The return per acre of this combination is therefore only £95 although the return per acre of coffee is £190. In a more sophisticated programme it might be possible to consider different coffee activities with different rates of mulching. Tea is treated alone as a single activity.

The return per unit of each activity generally corresponds with the gross margin; that is, the gross output minus variable costs. Thus seed, spray, fertilizer and manure costs are deducted from the crop gross outputs. Return per cow takes account of stock depreciation, the annual sale of a calf and spraying costs. It would appear that no depreciation charges are made against the tree crops. In fact, if the trees are already mature and yielding a crop then depreciation costs are unavoidable and may be ignored in the planning exercise. Where, however, the trees have not yet been planted, the costs of establishment and of waiting should be taken into account, preferably by discounting. However, this would clearly make the problem far more complicated. The costs of oxen hire, or tractor hire in an alternative programme, are treated as fixed costs not included in the linear programming exercise but deducted from the total return. However, these costs are specific to the arable crops and should therefore be taken into account in the

programming procedure either by treating them as a buying activity which relaxes a cultivation constraint on the arable crops or
by deducting a fixed cost from the individual arable crop returns.
Either of these approaches might reduce the area of arable crops
in the optimum solution.

Labour constraints are only included for certain months.
Earlier studies had shown that in the remaining months of May,
June, August and September, there is surplus labour. The labour
available to this holding was not known but the original budgeted
proposals required 3 full-time adults if oxen were used for seed-bed
cultivations. Assuming 300 working days annually, 3 men would
supply, on an 8-hour day basis, 600 man-hours a month. It was
therefore assumed that this is the monthly labour supply. The
arable crops naturally only require labour during the particular
season of growth. Similarly two land constraints were used, one
for the long rains and one for the short rains. Short rains crops do
not compete with long rains crops for land or labour. However,
leys and tree crops cover the ground continuously and require land
and labour in both seasons. The total acreage of the holding was
10·48 acres (4·30 hectares) but it was assumed that some of the
land would be retained in its current use for vegetables, bananas,
permanent grass and trees, so the area entering the programme
was reduced to 7·37 acres.

The rotational constraint required that four years of arable
cropping should be followed by three years of grass ley. There were
two technical reasons for this constraint, first that the land should
be rested periodically from arable cropping and second that sufficient dairy cows should be kept to provide farmyard manure for
the arable crops. Since grass leys in association with milk production are unlikely to enter a programme on their own merits, the
returns per unit of limiting resource being less than those from
perennial cash crops, and since they are a vital restriction from the
rotational point of view, an acceptable crop/ley ratio of 4:3 must
be built into the programme to ensure that the fertility requirements are met. This is achieved as follows:

Let $X_1 =$ acres (units) of English potatoes (long rains),
$\quad X_2 =$ acres of maize (long rains),
$\quad X_3 =$ acres of beans (long rains),
$\quad X_4 =$ acres of ley and milk.

Only long rains arable crops are considered, because the short
rains crops, if any, would be grown on the same arable land. The

constraint requires that the area of ley and milk (X_4) shall be at least three-quarters of the area of arable crops $(X_1, X_2$ and $X_3)$. Thus we can write:

$$X_4 \geqslant \tfrac{3}{4} (X_1 + X_2 + X_3).$$

Multiplying both sides by 4 and removing the brackets gives:

$$4X_4 \geqslant 3X_1 + 3X_2 + 3X_3.$$

Now moving the left-hand term to the right-hand side gives

$$0 \geqslant 3X_1 + 3X_2 + 3X_3 - 4X_4.$$

This is the form in which the constraint is expressed in the matrix. It can be readily converted into an equality by introducing a slack variable which allows the area of ley to exceed the minimum required.

The optimum solution obtained by linear programming these data is given in Table 16.11. This gives a net return of £389 in comparison with the £308 expected from the budgeted plan (see Table 14.1, page 273).

TABLE 16.11 Solution to linear programme of case study farm in Kenya

	Plan						
	LR potatoes	LR maize	LR beans	Ley and milk	Coffee and mulch	Tea	Total
Acres	1·21	0·65	0·45	1·73	1·26	2·07	7·37
Gross margin (£)	26·7	8·0	5·6	21·6	119·7	207·0	388·6

Disposal activities at positive level
    Labour man-days   October 200, November 190, January 108, February 87
    Land SR   2·31 acres (fallow)

Activities at zero level and respective Z value (£ per unit)

SR potatoes	SR maize	SR beans	March labour	April labour	July labour	December labour	LR land	Arable/ley ratio
25·1	22·2	11·1	0·03	0·12	0·04	0·45	2·0	2·0

Note. LR = long rains; SR = short rains.

A further $1\tfrac{1}{2}$ acres of permanent pasture is to be used for dairying and the return from this, together with the programmed return, less the cost of oxen hire gives a total net farm income of £404.

In practice slight modification is needed. The $1\frac{1}{2}$ acres of permanent pasture plus 1·73 acres of ley in the programmed solution would carry 3·23 cows if stocked at 1 cow to the acre. Of course, the figure must be rounded down to 3 cows and the expected income must be reduced accordingly.

The results suggest some interesting relationships between perennial cash crops and annual arable crops, which may be of relevance on other holdings in other situations. Firstly it is noteworthy that in the rotational area, the entire short rains acreage is left fallow. The implication is, of course, that short rains cropping, in association with cash-crop farming, is uneconomic. This not only goes against the universal practice of the Kenya farmer but contradicts the teaching of the agriculturalists as embodied in their farm plans. It is put forward as an example of the divergence between technical and economic efficiency. However, further work may be required to confirm this. The divergence could be due to mis-specification of the programming problem instead.

Another interesting feature is the acreage of cash crops in the plan, which is less than the total farm area. Many agriculturalists recommend planting the whole farm to cash crops in order to maximize profits. Cash crops do not occupy the entire acreage because labour is limiting and the long rains arable crops are supplementary to the cash crops in labour use. They require labour during the long rains whereas the main tree crop harvesting labour is needed at other times.

In considering this and, indeed, any other application of linear programming it is important to remember that the use of the computer does not improve the accuracy of the available data. If there are errors in the information which is used in linear programming then there will be errors in the solution obtained.

Because of the expense and time involved in linear and other forms of mathematical programming, these techniques are not suitable for application on individual farms.[10] However, there may be considerable scope for use of linear programming in solving generalized problems for advisory purposes. A linear programme prepared for a typical model farm can provide useful guidelines for offering advice on a large number of similar farms. Clearly this kind of work would have to be carried out by research staff working

---

[10] Many new mathematical techniques involving the use of computers have been used in studying farm management problems in recent years. Apart from linear programming, however, they are not in widespread use in Africa.

in cooperation with the advisory services. Similarly, linear programming may be used, indeed has been used, for planning large irrigation and other settlement schemes.

SUGGESTIONS FOR FURTHER READING

BOUSSARD, J. M. and PETIT, M. (1967) Representation of farmers' behaviour under uncertainty with a focus-loss constraint, *American Journal of Agricultural Economics*, vol. 49, no. 4, p. 869.

CLAYTON, E. S. (1963) *Economic planning in peasant agriculture*, University of London, Wye College, Department of Agricultural Economics.

DENT, J. B. and CASEY, H. (1967) *Linear programming and animal nutrition*, London, Crosby Lockwood.

HEADY, E. O. and CANDLER, W. (1958) *Linear programming methods*, Iowa State University Press.

OGUNFOWORA, O. (1970) Optimum farm plans for arable-crop farm settlement in Western Nigeria: a polyperiod linear programming analysis, *Nigerian Journal of Economic and Social Studies*, vol. 12, no. 2, p. 205.

STEWART, J. D. (1961) Farm operating capital as a constraint: a problem on the application of linear programming, *Farm Economist*, vol. 9, no. 10, p. 463.

THROSBY, C. D. (1970) *Elementary linear programming*, New York, Random House.

HEYER, J. (1971) A linear programming analysis of constraints on peasant farms in Kenya, *Food Research Institute Studies*, vol. 10, no. 1, p. 55.

Appendix

TABLE I Future value (V_n) of £1 in n years. $V_n = 1(1 + r)^n$

Rate of interest (r) per cent

	2·00	4·00	6·00	8·00	10·00	12·00	14·00	16·00	18·00	20·00	22·00	24·00	26·00	28·00	30·00
1	1·02	1·04	1·06	1·08	1·10	1·12	1·14	1·16	1·18	1·20	1·22	1·24	1·26	1·28	1·30
2	1·04	1·08	1·12	1·17	1·21	1·25	1·30	1·35	1·39	1·44	1·49	1·54	1·59	1·64	1·69
3	1·06	1·12	1·19	1·26	1·33	1·40	1·48	1·56	1·64	1·73	1·82	1·91	2·00	2·10	2·20
4	1·08	1·17	1·26	1·36	1·46	1·57	1·69	1·81	1·94	2·07	2·22	2·36	2·52	2·68	2·86
5	1·10	1·22	1·34	1·47	1·61	1·76	1·93	2·10	2·29	2·49	2·70	2·93	3·18	3·44	3·71
6	1·13	1·27	1·42	1·59	1·77	1·97	2·19	2·44	2·70	2·99	3·30	3·64	4·00	4·40	4·83
7	1·15	1·32	1·50	1·71	1·95	2·21	2·50	2·83	3·19	3·58	4·02	4·51	5·04	5·63	6·27
8	1·17	1·37	1·59	1·85	2·14	2·48	2·85	3·28	3·76	4·30	4·91	5·59	6·35	7·21	8·16
9	1·20	1·42	1·69	2·00	2·36	2·77	3·25	3·80	4·44	5·16	5·99	6·93	8·00	9·22	10·60
10	1·22	1·48	1·79	2·16	2·59	3·11	3·71	4·41	5·23	6·19	7·30	8·59	10·09	11·81	13·79
11	1·24	1·54	1·90	2·33	2·85	3·48	4·23	5·12	6·18	7·43	8·91	10·66	12·71	15·11	17·92
12	1·27	1·60	2·01	2·52	3·14	3·90	4·82	5·94	7·29	8·92	10·87	13·21	16·01	19·34	23·30
13	1·29	1·67	2·13	2·72	3·45	4·36	5·49	6·89	8·60	10·70	13·26	16·39	20·18	24·76	30·29
14	1·32	1·73	2·26	2·94	3·80	4·89	6·26	7·99	10·15	12·84	16·18	20·32	25·42	31·69	39·37
15	1·35	1·80	2·40	3·17	4·18	5·47	7·14	9·27	11·97	15·41	19·74	25·20	32·03	40·56	51·19
16	1·37	1·87	2·54	3·43	4·59	6·13	8·14	10·75	14·13	18·49	24·09	31·24	40·36	51·92	66·54
17	1·40	1·95	2·69	3·70	5·05	6·87	9·28	12·47	16·67	22·19	29·38	38·74	50·85	66·46	86·50
18	1·43	2·03	2·85	4·00	5·56	7·69	10·58	14·46	19·67	26·62	35·85	48·04	64·07	85·07	112·46
19	1·46	2·11	3·03	4·32	6·12	8·61	12·06	16·78	23·21	31·95	43·74	59·57	80·73	108·89	146·19
20	1·49	2·19	3·21	4·66	6·73	9·65	13·74	19·46	27·39	38·34	53·36	73·86	101·72	139·38	190·05
21	1·52	2·28	3·40	5·03	7·40	10·80	15·67	22·57	32·32	46·01	65·10	91·59	128·17	178·41	247·06
22	1·55	2·37	3·60	5·44	8·14	12·10	17·86	26·19	38·14	55·21	79·42	113·57	161·49	228·36	321·18
23	1·58	2·46	3·82	5·87	8·95	13·55	20·36	30·38	45·01	66·25	96·89	140·83	203·48	292·30	417·54
24	1·61	2·56	4·05	6·34	9·85	15·18	23·21	35·24	53·11	79·50	118·21	174·63	256·39	374·14	542·80
25	1·64	2·67	4·29	6·85	10·83	17·00	26·46	40·87	62·67	95·40	144·21	216·54	323·05	478·90	705·64
26	1·67	2·77	4·55	7·40	11·92	19·04	30·17	47·41	73·95	114·48	175·94	268·51	407·04	613·00	917·33
27	1·71	2·88	4·82	7·99	13·11	21·32	34·39	55·00	87·26	137·37	214·64	332·95	512·87	784·64	1192·53
28	1·74	3·00	5·11	8·63	14·42	23·88	39·20	63·80	102·97	164·84	261·86	412·86	646·21	1004·34	1550·29
29	1·78	3·12	5·42	9·32	15·86	26·75	44·69	74·01	121·50	197·81	319·47	511·95	814·23	1285·55	2015·38
30	1·81	3·24	5·74	10·06	17·45	29·96	50·95	85·85	143·37	237·38	389·76	634·82	1025·93	1645·50	2620·00

Note. More detailed figures are given in Lawson, G. H. and Windle, D. W. (1965) *Tables or discounted cash flow, annuity, sinking fund, compound interest and annual capital charge calculations*, Edinburgh, Oliver and Boyd.

TABLE II Future value (V_{an}) at n years of an annuity of £1. $V_{an} = \dfrac{1[(1 + r)^n - 1]}{r}$

Rate of interest (r) per cent

	2·00	4·00	6·00	8·00	10·00	12·00	14·00	16·00	18·00	20·00	22·00	24·00	26·00	28·00	30·00
1	1·00	1·00	1·00	1·00	1·00	1·00	1·00	1·00	1·00	1·00	1·00	1·00	1·00	1·00	1·00
2	2·02	2·04	2·06	2·08	2·10	2·12	2·14	2·16	2·18	2·20	2·22	2·24	2·26	2·28	2·30
3	3·06	3·12	3·18	3·25	3·31	3·37	3·44	3·51	3·57	3·64	3·71	3·78	3·85	3·92	3·99
4	4·12	4·25	4·37	4·51	4·64	4·78	4·92	5·07	5·22	5·37	5·52	5·68	5·85	6·02	6·19
5	5·20	5·42	5·64	5·87	6·11	6·35	6·61	6·88	7·15	7·44	7·74	8·05	8·37	8·70	9·04
6	6·31	6·63	6·98	7·34	7·72	8·12	8·54	8·98	9·44	9·93	10·44	10·98	11·54	12·14	12·76
7	7·43	7·90	8·39	8·92	9·49	10·09	10·73	11·41	12·14	12·92	13·74	14·62	15·55	16·53	17·58
8	8·58	9·21	9·90	10·64	11·44	12·30	13·23	14·24	15·33	16·50	17·76	19·12	20·59	22·16	23·86
9	9·75	10·58	11·49	12·49	13·58	14·78	16·09	17·52	19·09	20·80	22·67	24·71	26·94	29·37	32·01
10	10·95	12·01	13·18	14·49	15·94	17·55	19·34	21·32	23·52	25·96	28·66	31·64	34·94	38·59	42·62
11	12·17	13·49	14·97	16·65	18·53	20·65	23·04	25·73	28·76	32·15	35·96	40·24	45·03	50·40	56·41
12	13·41	15·03	16·87	18·98	21·38	24·13	27·27	30·85	34·93	39·58	44·87	50·89	57·74	65·51	74·33
13	14·68	16·63	18·88	21·50	24·52	28·03	32·09	36·79	42·22	48·50	55·75	64·11	73·75	84·85	97·63
14	15·97	18·29	21·02	24·21	27·97	32·39	37·58	43·67	50·82	59·20	69·01	80·50	93·93	109·61	127·91
15	17·29	20·02	23·28	27·15	31·77	37·28	43·84	51·66	60·97	72·04	85·19	100·82	119·35	141·30	167·29
16	18·64	21·82	25·67	30·32	35·95	42·75	50·98	60·93	72·94	87·44	104·93	126·01	151·38	181·87	218·47
17	20·01	23·70	28·21	33·75	40·55	48·88	59·12	71·67	87·07	105·93	129·02	157·25	191·73	233·79	285·01
18	21·41	25·65	30·91	37·45	45·60	55·75	68·39	84·14	103·74	128·12	158·40	195·99	242·59	300·25	371·52
19	22·84	27·67	33·76	41·45	51·16	63·44	78·97	98·60	123·41	154·74	194·25	244·03	306·66	385·32	483·97
20	24·30	29·78	36·79	45·76	57·27	72·05	91·02	115·38	146·63	186·69	237·99	303·60	387·39	494·21	630·17
21	25·78	31·97	39·99	50·42	64·00	81·70	104·77	134·84	174·02	225·03	291·35	377·46	489·11	633·59	820·22
22	27·30	34·25	43·39	55·46	71·40	92·50	120·44	157·41	206·34	271·03	356·44	469·06	617·28	812·00	1067·28
23	28·84	36·62	47·00	60·89	79·54	104·60	138·30	183·60	244·49	326·24	435·86	582·63	778·77	1040·36	1388·46
24	30·42	39·08	50·82	66·76	88·50	118·16	158·66	213·98	289·49	392·48	532·75	723·46	982·25	1332·66	1806·00
25	32·03	41·65	54·86	73·11	98·35	133·33	181·87	249·21	342·60	471·98	650·96	898·09	1238·64	1706·80	2348·80
26	33·67	44·31	59·16	79·95	109·18	150·33	208·33	290·09	405·27	567·38	795·17	1114·63	1561·68	2185·71	3054·44
27	35·34	47·08	63·71	87·35	121·10	169·37	238·50	337·50	479·22	681·85	971·10	1383·15	1968·72	2798·71	3971·78
28	37·05	49·97	68·53	95·34	134·21	190·70	272·89	392·50	566·48	819·22	1185·74	1716·10	2481·59	3583·34	5104·31
29	38·79	52·97	73·64	103·97	148·63	214·58	312·09	456·30	669·45	984·07	1447·61	2128·96	3127·80	4587·68	6714·60
30	40·57	56·08	79·06	113·28	164·49	241·33	356·79	530·31	790·95	1181·88	1767·08	2640·92	3942·03	5873·23	8729·99

Table III 333

TABLE III Present value (P) of £1 due in n years. $P = \dfrac{1}{(1+r)^n}$

Rate of interest (r) per cent

n	2.00	4.00	6.00	8.00	10.00	12.00	14.00	16.00	18.00	20.00	22.00	24.00	26.00	28.00	30.00
1	0.9804	0.9615	0.9434	0.9259	0.9091	0.8929	0.8772	0.8621	0.8475	0.8333	0.8197	0.8065	0.7937	0.7813	0.7692
2	0.9612	0.9246	0.8900	0.8573	0.8264	0.7972	0.7695	0.7432	0.7182	0.6944	0.6719	0.6504	0.6299	0.6104	0.5917
3	0.9423	0.8890	0.8396	0.7938	0.7513	0.7118	0.6750	0.6407	0.6086	0.5787	0.5507	0.5245	0.4999	0.4768	0.4552
4	0.9238	0.8548	0.7921	0.7350	0.6830	0.6355	0.5921	0.5523	0.5158	0.4823	0.4514	0.4230	0.3968	0.3725	0.3501
5	0.9057	0.8219	0.7473	0.6806	0.6209	0.5674	0.5194	0.4761	0.4371	0.4019	0.3700	0.3411	0.3149	0.2910	0.2693
6	0.8880	0.7903	0.7050	0.6302	0.5645	0.5066	0.4556	0.4104	0.3704	0.3349	0.3033	0.2751	0.2499	0.2274	0.2072
7	0.8706	0.7599	0.6651	0.5835	0.5132	0.4523	0.3996	0.3538	0.3139	0.2791	0.2486	0.2218	0.1983	0.1776	0.1594
8	0.8535	0.7307	0.6274	0.5403	0.4665	0.4039	0.3506	0.3050	0.2660	0.2326	0.2038	0.1789	0.1574	0.1388	0.1226
9	0.8368	0.7026	0.5919	0.5002	0.4241	0.3606	0.3075	0.2630	0.2255	0.1938	0.1670	0.1443	0.1249	0.1084	0.0943
10	0.8203	0.6756	0.5584	0.4632	0.3855	0.3220	0.2697	0.2267	0.1911	0.1615	0.1369	0.1164	0.0992	0.0847	0.0725
11	0.8043	0.6496	0.5268	0.4289	0.3505	0.2875	0.2366	0.1954	0.1619	0.1346	0.1122	0.0938	0.0787	0.0662	0.0558
12	0.7885	0.6246	0.4970	0.3971	0.3186	0.2567	0.2076	0.1685	0.1372	0.1122	0.0920	0.0757	0.0625	0.0517	0.0429
13	0.7730	0.6006	0.4688	0.3677	0.2897	0.2292	0.1821	0.1452	0.1163	0.0935	0.0754	0.0610	0.0496	0.0404	0.0330
14	0.7579	0.5775	0.4423	0.3405	0.2633	0.2046	0.1597	0.1252	0.0985	0.0779	0.0618	0.0492	0.0393	0.0316	0.0254
15	0.7430	0.5553	0.4173	0.3152	0.2394	0.1827	0.1401	0.1079	0.0835	0.0649	0.0507	0.0397	0.0312	0.0247	0.0195
16	0.7284	0.5339	0.3936	0.2919	0.2176	0.1631	0.1229	0.0930	0.0708	0.0541	0.0415	0.0320	0.0248	0.0193	0.0150
17	0.7142	0.5134	0.3714	0.2703	0.1978	0.1456	0.1078	0.0802	0.0600	0.0451	0.0340	0.0258	0.0197	0.0150	0.0116
18	0.7002	0.4936	0.3503	0.2502	0.1799	0.1300	0.0946	0.0691	0.0508	0.0376	0.0279	0.0208	0.0156	0.0118	0.0089
19	0.6864	0.4746	0.3305	0.2317	0.1635	0.1161	0.0829	0.0596	0.0431	0.0313	0.0229	0.0168	0.0124	0.0092	0.0068
20	0.6730	0.4564	0.3118	0.2145	0.1486	0.1037	0.0728	0.0514	0.0365	0.0261	0.0187	0.0135	0.0098	0.0072	0.0053
21	0.6598	0.4388	0.2942	0.1987	0.1351	0.0926	0.0638	0.0443	0.0309	0.0217	0.0154	0.0109	0.0078	0.0056	0.0040
22	0.6468	0.4220	0.2775	0.1839	0.1228	0.0826	0.0560	0.0382	0.0262	0.0181	0.0126	0.0088	0.0062	0.0044	0.0031
23	0.6342	0.4057	0.2618	0.1703	0.1117	0.0738	0.0491	0.0329	0.0222	0.0151	0.0103	0.0071	0.0049	0.0034	0.0024
24	0.6217	0.3901	0.2470	0.1577	0.1015	0.0659	0.0431	0.0284	0.0188	0.0126	0.0085	0.0057	0.0039	0.0027	0.0018
25	0.6095	0.3751	0.2330	0.1460	0.0923	0.0588	0.0378	0.0245	0.0160	0.0105	0.0069	0.0046	0.0031	0.0021	0.0014
26	0.5976	0.3607	0.2198	0.1352	0.0839	0.0525	0.0331	0.0211	0.0135	0.0087	0.0057	0.0037	0.0025	0.0016	0.0011
27	0.5859	0.3468	0.2074	0.1252	0.0763	0.0469	0.0291	0.0182	0.0115	0.0073	0.0047	0.0030	0.0019	0.0013	0.0008
28	0.5744	0.3335	0.1956	0.1159	0.0693	0.0419	0.0255	0.0157	0.0097	0.0061	0.0038	0.0024	0.0015	0.0010	0.0006
29	0.5631	0.3207	0.1846	0.1073	0.0630	0.0374	0.0224	0.0135	0.0082	0.0051	0.0031	0.0020	0.0012	0.0008	0.0005
30	0.5521	0.3083	0.1741	0.0994	0.0573	0.0334	0.0196	0.0116	0.0070	0.0042	0.0026	0.0016	0.0010	0.0006	0.0004

T<small>ABLE</small> IV Present value (P_{an}) of an annuity of £1 payable for n future years. $P_{an} = \dfrac{1[(1+r)^n - 1]}{r(1+r)^n}$

Rate of interest (r) per cent

n	2·00	4·00	6·00	8·00	10·00	12·00	14·00	16·00	18·00	20·00	22·00	24·00	26·00	28·00	30·00
1	0·9804	0·9615	0·9434	0·9259	0·9091	0·8929	0·8772	0·8621	0·8475	0·8333	0·8197	0·8065	0·7937	0·7813	0·7692
2	1·9416	1·8861	1·8334	1·7833	1·7355	1·6901	1·6467	1·6052	1·5556	1·5278	1·4915	1·4568	1·4235	1·3916	1·3609
3	2·8839	2·7751	2·6730	2·5771	2·4869	2·4018	2·3216	2·2459	2·1743	2·1065	2·0422	1·9813	1·9234	1·8684	1·8161
4	3·8077	3·6299	3·4651	3·3121	3·1699	3·0373	2·9137	2·7982	2·6901	2·5887	2·4936	2·4043	2·3202	2·2410	2·1662
5	4·7135	4·4518	4·2124	3·9927	3·7908	3·6048	3·4331	3·2743	3·1272	2·9906	2·8636	2·7454	2·6351	2·5320	2·4356
6	5·6014	5·2421	4·9173	4·6229	4·3553	4·1114	3·8887	3·6847	3·4976	3·3255	3·1669	3·0205	2·8850	2·7594	2·6427
7	6·4720	6·0021	5·5824	5·2064	4·8684	4·5638	4·2883	4·0386	3·8115	3·6046	3·4155	3·2423	3·0833	2·9370	2·8021
8	7·3255	6·7327	6·2098	5·7466	5·3349	4·9676	4·6389	4·3436	4·0776	3·8372	3·6193	3·4212	3·2407	3·0758	2·9247
9	8·1622	7·4353	6·8017	6·2469	5·7590	5·3282	4·9464	4·6065	4·3030	4·0310	3·7863	3·5655	3·3657	3·1842	3·0190
10	8·9826	8·1109	7·3601	6·7101	6·1446	5·6502	5·2161	4·8332	4·4941	4·1925	3·9232	3·6819	3·4648	3·2689	3·0915
11	9·7868	8·7605	7·8869	7·1390	6·4951	5·9377	5·4527	5·0286	4·6560	4·3271	4·0354	3·7757	3·5435	3·3351	3·1473
12	10·5753	9·3851	8·3838	7·5361	6·8137	6·1944	5·6603	5·1971	4·7932	4·4392	4·1274	3·8514	3·6059	3·3868	3·1903
13	11·3484	9·9856	8·8527	7·9038	7·1034	6·4235	5·8424	5·3423	4·9095	4·5327	4·2028	3·9124	3·6555	3·4272	3·2233
14	12·1062	10·5631	9·2950	8·2442	7·3667	6·6282	6·0021	5·4675	5·0081	4·6106	4·2646	3·9616	3·6949	3·4587	3·2487
15	12·8493	11·1184	9·7122	8·5595	7·6061	6·8109	6·1422	5·5755	5·0916	4·6755	4·3152	4·0013	3·7261	3·4834	3·2682
16	13·5777	11·6523	10·1059	8·8514	8·0237	6·9740	6·2651	5·6685	5·1624	4·7296	4·3567	4·0333	3·7509	3·5026	3·2832
17	14·2919	12·1657	10·4773	9·1216	8·0216	7·1196	6·3729	5·7487	5·2223	4·7746	4·3908	4·0591	3·7705	3·5177	3·2948
18	14·9920	12·6593	10·8276	9·3719	8·2014	7·2497	6·4674	5·8178	5·2732	4·8122	4·4187	4·0799	3·7861	3·5294	3·3037
19	15·6785	13·1339	11·1581	9·6036	8·3649	7·3658	6·5504	5·8775	5·3162	4·8435	4·4415	4·0967	3·7985	3·5386	3·3105
20	16·3514	13·5903	11·4699	9·8181	8·5136	7·4694	6·6231	5·9288	5·3527	4·8696	4·4603	4·1103	3·8083	3·5458	3·3158
21	17·0112	14·0292	11·7641	10·0168	8·6487	7·5620	6·6870	5·9731	5·3837	4·8913	4·4756	4·1212	3·8161	3·5514	3·3198
22	17·6580	14·4511	12·0416	10·2007	8·7715	7·6446	6·7429	6·0113	5·4099	4·9094	4·4882	4·1300	3·8223	3·5558	3·3230
23	18·2922	14·8568	12·3034	10·3711	8·8832	7·7184	6·7921	6·0442	5·4321	4·9245	4·4985	4·1371	3·8273	3·5592	3·3254
24	18·9139	15·2470	12·5504	10·5288	8·9847	7·7843	6·8351	6·0726	5·4509	4·9371	4·5070	4·1428	3·8312	3·5619	3·3272
25	19·5235	15·6221	12·7834	10·6748	9·0770	7·8431	6·8729	6·0971	5·4669	4·9476	4·5139	4·1474	3·8342	3·5640	3·3286
26	20·1210	15·9828	13·0032	10·8100	9·1609	7·8957	6·9061	6·1182	5·4804	4·9563	4·5196	4·1511	3·8367	3·5656	3·3297
27	20·7069	16·3296	13·2105	10·9352	9·2372	7·9426	6·9352	6·1364	5·4919	4·9636	4·5243	4·1542	3·8387	3·5669	3·3305
28	21·2813	16·6631	13·4062	11·0511	9·3066	7·9844	6·9607	6·1520	5·5016	4·9697	4·5281	4·1566	3·8402	3·5679	3·3312
29	21·8444	16·9837	13·5907	11·1584	9·3696	8·0218	6·9830	6·1656	5·5098	4·9747	4·5312	4·1585	3·8414	3·5687	3·3317
30	22·3965	17·2920	13·7648	11·2578	9·4269	8·0552	7·0027	6·1772	5·5168	4·9789	4·5338	4·1601	3·8424	3·5693	3·3321

Index